DISASTERS OF ONONDAGA COUNTY

DISASTERS OF ONONDAGA COUNTY

NEIL K. MACMILLAN

THE History PRESS

Published by The History Press
Charleston, SC
www.historypress.net

First published 2017

Manufactured in the United States

ISBN 9781467137867

Library of Congress Control Number: 2017934927

CONTENTS

For Peggy.

ACKNOWLEDGEMENTS

I'd like to thank the staff at the Onondaga Historical Association, the Post-Standard Newspaper Archives and everyone on the Facebook page Nostalgic Syracuse. You all contributed useful ideas, and several of you included links that saved me hours of searching. Special thanks to John Blair for sharing his expertise in railroad and trolley history for the area.

1

AN OVERVIEW OF ONONDAGA COUNTY AND OUR DISASTERS

The news is full of disaster. Turn on the television or surf the web and you'll find stories of man-made or natural mayhem. Usually we're spared these ravages. Syracuse and Onondaga County don't face a lot of calamity from Mother Nature. In fact, Syracuse was named the safest city in the United States for avoiding natural disasters.[1] As those of us who live here know, we're prone to get snow measured in feet.[2] While it is highly unusual to get tornados, hurricanes and earthquakes in Onondaga County, it is not impossible. Indeed, Longbranch Park—which is a bucolic picnic venue these days—was devastated by a tornado in 1912. Back then, it was a resort and amusement park. Mild earthquakes have gently nudged us, just to ensure we don't get too complacent, and we have suffered through tropical storms, notably Hurricane Hazel in 1954, but these are the exception rather than the rule for Onondaga County.

Let's face it: we are our own worst enemies. Before you pillory me, look at the majority of disasters that have happened over the years in Onondaga County. There have been train, plane, bus, trolley and automobile accidents. Fires ravaged the city, notably the Bastable fire of 1923 and the Collins Block fire that claimed the lives of eight firefighters in 1939. There have been explosions of gunpowder, munitions and steam. If we can build it, we can destroy it in a spectacular fashion or watch it fall down in equally stunning fashion. The Greenway Brewery suffered a floor collapse in 1880. The Central Baptist Church endured the same fate six years earlier.

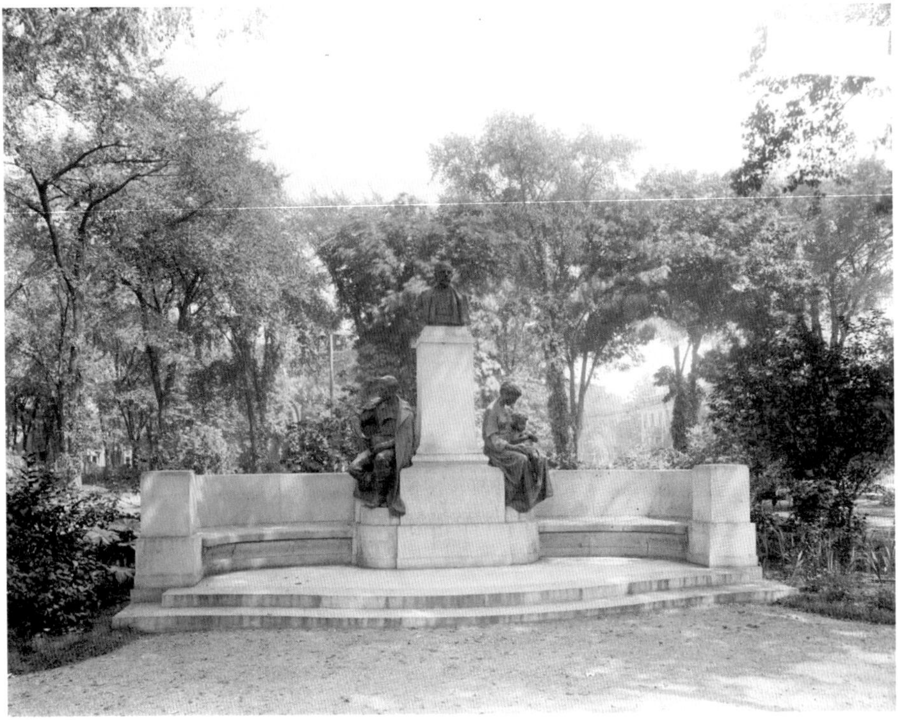

Fire gutted the Granger Block in Syracuse. *Courtesy of the Library of Congress.*

Briefly put, a disaster is a calamity that brings great destruction or loss of life. At least, that is how the dictionary defines it.[3] Things have gone horribly wrong in a hurry. Most of the incidents in this book were accidental. The notable exception was the bombing of Pan-Am Flight 103 over Lockerbie, Scotland, which cost several Syracuse University students their lives on December 21, 1988.

Disease raced through the county. Early in the county's history, two soldiers succumbed to smallpox during the War of 1812. They are buried in a small plot in Onondaga Hill. Smallpox would ravage Syracuse sixty-three years later. We were affected by the Spanish flu pandemic of 1918. We fought cholera, malaria and measles. Much of the devastation was visited on the county thanks to polluted water and lack of sanitary practices.

The story of disaster isn't only about what happened. That is but a small part. The events I write about are more about why they happened and how we reacted. There are heroes and villains here. Some of them are in the open.

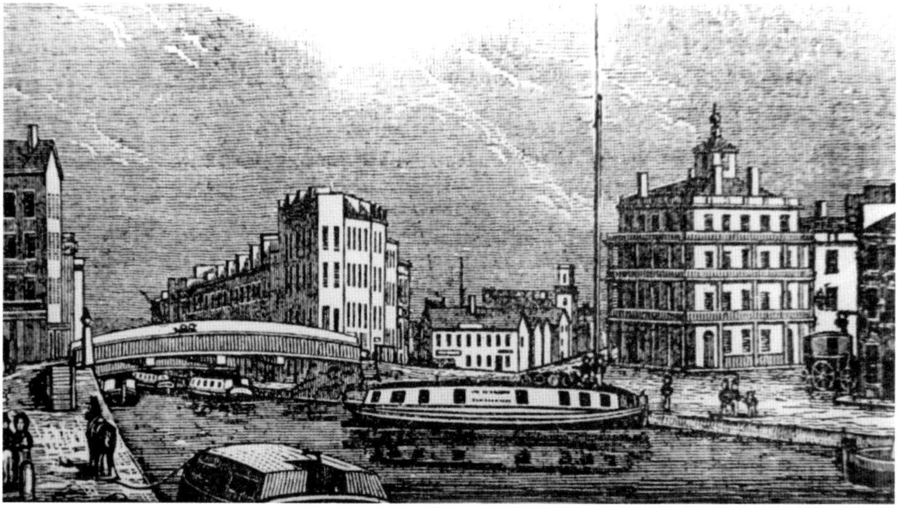

A lithograph of Clinton Square in the 1840s. Much has changed in 170 years. *Courtesy of the Library of Congress.*

Others hide in the shadows. It takes a special kind of person to rush into a burning building at extreme peril to themselves to save people they usually don't know. In Syracuse, there is a professional fire department, but in most of the county and for a good portion of Syracuse's history, the firefighters were volunteers. These men and women put their lives on the line to save our lives and property and, in the case of volunteer fire departments, do so without pay. Then there are the unsung heroes and heroines who go out of their way to help, never expecting anything. They step up when a need arises and then return to anonymity. Many of the villains remain nameless. Only one person was ever indicted for the terrorist attack on Pam-Am 103. Others have been seen as both, sometimes at the same time.

As you have heard me say before, history is a journey. Our journey will take us back through the mists of time, but we won't be going far from home; all that makes us collectively Onondaga County can be found in our history. Bring a jacket—our trek will encounter inclement weather. You might want good shoes and a Swiss Army knife, too. Step softly, for we're entering places that serve as final resting places even if there are no headstones.

2

WHEN THE WIND BLOWS

Hurricanes, Tornadoes and a Derecho

Storms were the first disasters faced by mankind. Before there were cities or civilization of any kind, there was storm. Wind was both worshiped and feared by our primitive ancestors. Prayers and sacrifices were made to keep evil forces in the guise of storms at bay and to hopefully entice good forces and beneficial weather. Storms could starve nomadic people by driving away the game they depended on and also kill or injure members of the hunter-gatherer bands we all descend from. Some of the most horrifying and devastating carnage Mother Nature can wreak on us come from the wind. I don't know if parents sing the old lullaby "Rock a Bye Baby" to their toddlers anymore,[4] but from our earliest years, we are conditioned to fear high wind. Sometimes we're not wrong.

In this day of instant access to news anytime and anywhere, we can see the awesome fury hurricanes, tornadoes and microburst storms unleash. Usually, here in Onondaga County we only see this via the magic of television. Lest we get complacent, there have been notable windstorms in our history and in the not so distant past. When L. Frank Baum wrote *The Wizard of Oz*, a Kansas tornado was a central plot device to the story.[5] The tornado in Baum's children's book was based on an event he lived through on the plains of the Jayhawk State. We don't think of tornadoes when we think of Central New York, yet they do happen here.

In 1912, Woodrow Wilson was elected president, defeating a Republican Party split between William Howard Taft, the sitting president, and his mentor, former president Theodore Roosevelt. Onondaga County was prosperous due to fifty years of industrial growth. The eastern shore of Onondaga Lake was lined with resorts and amusement parks. One of the most popular was Long Branch Park, near the village of Liverpool. Trolley lines ran excursions to the park, and companies took employees there for company picnics. The park and its rivals along the shores of the lake were places of gaiety and merriment in the second decade of the twentieth century. Sunday, September 15, 1912, promised more of the same.

The tornado struck as clocks chimed five o'clock in the evening. The tornado was the first ever recorded in Onondaga County's history. The storm cut a swath through Onondaga County, from Onondaga Lake's western shore to South Bay on the southern shore of Oneida Lake. Three people were killed and forty injured. The tornado caused $250,000 in damages. Seventy-eight buildings were damaged. Forty of those were completely destroyed, but it could have been so much worse. The tornado's path missed the city of Syracuse by a few miles. Had the storm struck downtown Syracuse, the monetary damages would have been in the millions of dollars and the casualty figures gruesome.

Charles Dopp was cleaning the windshield of his trolley car when the tornado struck. He and the car were picked up by the howling wind and slammed into one of the many buildings wrecked by the tornado. He was taken to Genesee Hospital and died there of his injuries. William Madison and his wife were returning home to their farm in Longbranch and had just descended from the trolley at Longbranch Station when the storm struck.[6] Mr. Madison was pummeled by debris from the smashed station and killed instantly. Mrs. Madison's spine was broken in two places. Charles Chapman died after suffering internal injuries when his house and store stuck in the hamlet of Pitcher Hill in the town of Salina were struck. He was buried under debris and died of his injuries at St. Joseph's Hospital.

Syracuse and the county pulled together to raise money to help those whose lives were turned upside down by the tornado of 1912. The resorts and amusement parks are gone now. Changing tastes in entertainment and increasing pollution in the lake rendered them obsolete. Onondaga Lake is recovering from years of industrial abuses. The DestinyUSA mall fills some of the void left when the amusement parks disappeared, and the lake is clean enough—after years of intense labor and millions of dollars spent—to fish in again. Longbranch Park offers a pastoral setting for picnickers and those seeking some relaxing time away from the weekly grind.

Generally, Onondaga County is immune from the ravages of hurricanes and tropical storms. Usually, they break up or lose power before they reach

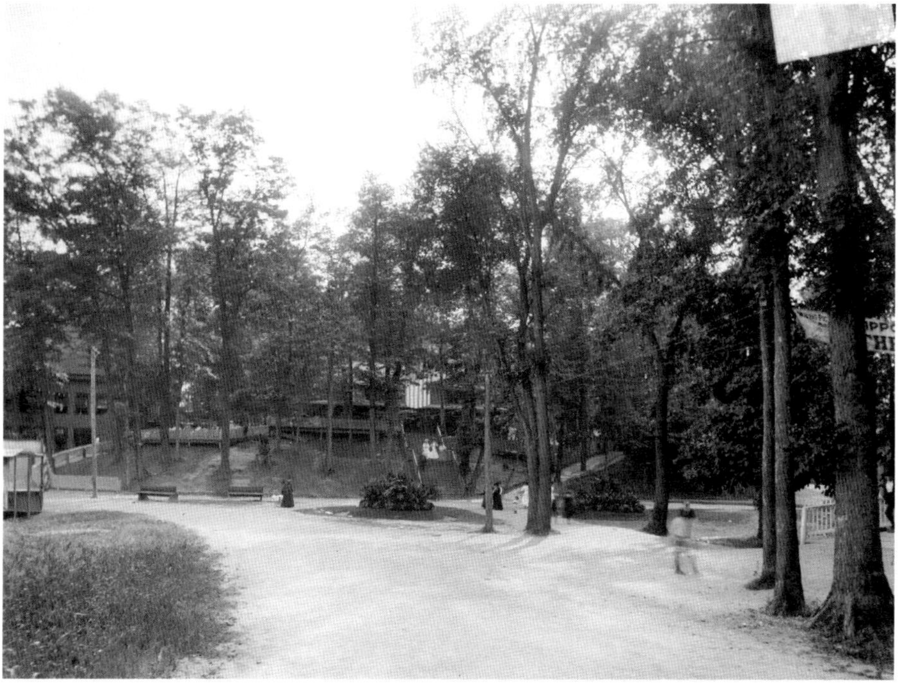

The tornado of 1912 cut a swath of destruction through the county, leaving three dead and this ruin in North Syracuse. *Courtesy of the Library of Congress.*

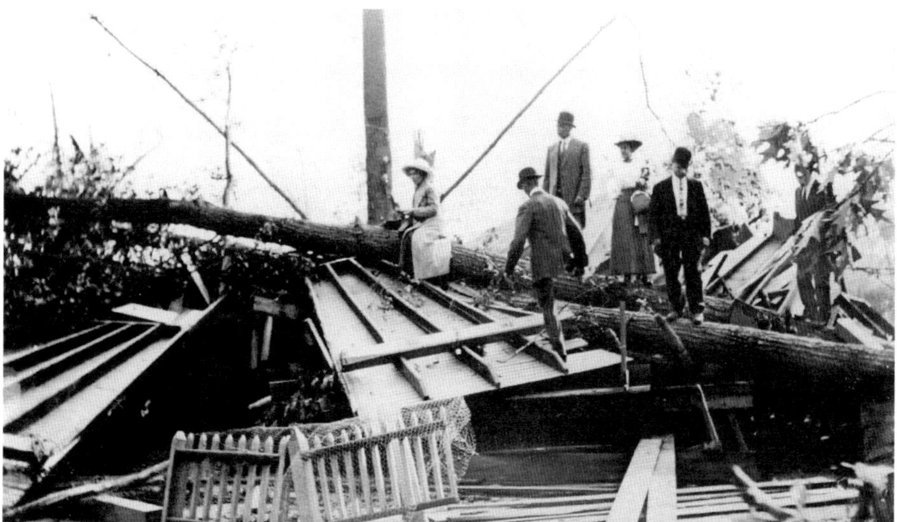

Onlookers survey the damage wrought by the 1912 tornado. *Courtesy of the Liverpool Public Library, from the Schuelke Collection.*

us. We see effects of the storms in copious amounts of rain for a few days, but we rarely get the winds that result in the massive devastation seen in the coastal regions of the nation. Syracusans tell a joke that addresses this: why did Florida get hurricanes and Syracuse get blizzards? Florida got first choice. It's not a terribly funny joke and isn't entirely accurate.

I grew up listening to my parents and maternal grandparents telling stories of Hurricane Hazel striking the area. Hurricane Hazel wreaked havoc from Haiti to Ontario, Canada. It was the seventeenth deadliest in the United States.[7] It was also the fifteenth costliest, causing $281 million in damages in 1954, an amount that was the equivalent of $1.057 billion in 1980. At 938 millibars and 27.7 inches mercury as a category 4 storm on the Saffir-Simpson scale, it was the eleventh-most-intense hurricane in the twentieth century up to 1983.[8] The village of Skaneateles was hardest hit by Hazel, but no area of Onondaga County was spared the hurricane's attention. The first casualty from Syracuse was an air crewman photographing the storm from a navy hurricane hunter aircraft. The

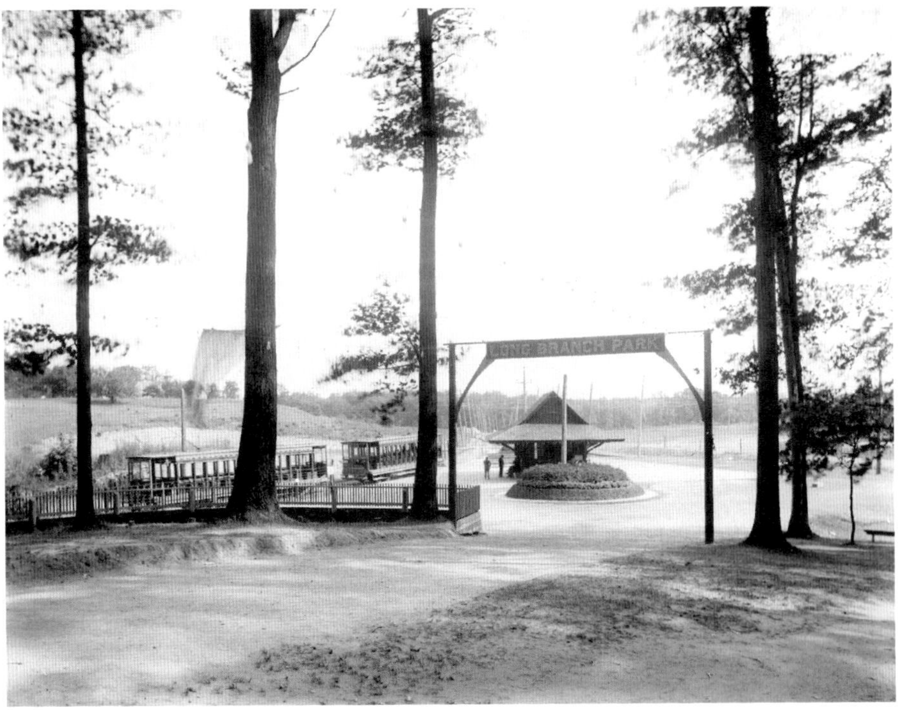

In its heyday, Longbranch Park was a major attraction for Onondaga County residents. *Courtesy of the Library of Congress.*

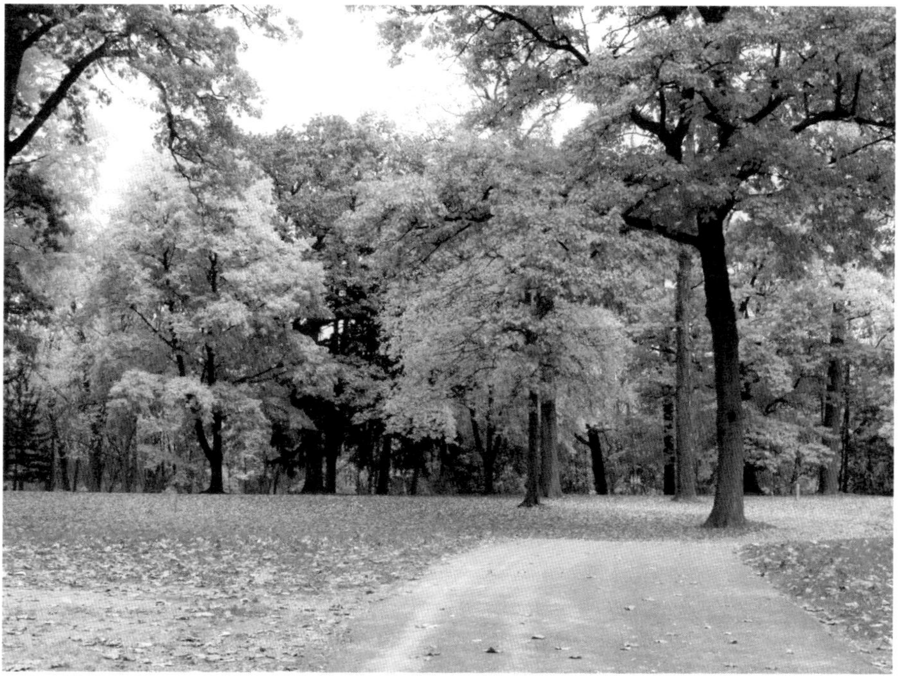

Today, Longbranch Park is a bucolic and popular picnic and recreation site for Onondaga County residents. *Neil K. MacMillan.*

aircraft was buffeted by the storm winds, and the sailor dislocated his shoulder. He was treated at Guantanamo Naval Station and returned to his base in Jacksonville, Florida. The storm was so devastating that the name Hazel was retired. There will never be another hurricane bearing that name.[9] Twelve people were killed in Upstate New York. None was from Onondaga County. The county did not escape Hazel's wrath, however. Thousands of trees were felled by the storm, cars were crushed and houses damaged. A Will and Baumer Candle company warehouse suffered damage when a wall was blown out by the hurricane.[10] Highways and streets were blocked throughout the area.

The storm that struck Onondaga County was an offshoot of Hurricane Hazel and is recognized as part of the main hurricane. In this day of instant information, we seem to not listen to the stories of our elders. When I was growing up, Hurricane Hazel was mythic. My parents and grandparents related stories of the storm's power and fury. We have seen and will see more storms. I don't know that we'll see another storm like Hazel. In these days of climate change and El Niño it is certainly possible. Will people be talking about them and tell tales about them decades later? I hope so.

Let's face it: wind can be the most vicious and devastating force on the planet. It is no respecter of property or the time spent building your life. Labor Day is the unofficial end of summer here in Upstate New York. When we think of destructive winds, it is the cyclonic monsters that we worry about. History and folklore abound with tales of tornadoes and hurricanes. For children of my generation, our first exposure to destructive winds may well have been watching *The Wizard of Oz* on television. Imagine then a straight line of wind packing the wallop of a tornado.

Labor Day weekend 1998 was full of promise for Central New York. The New York State fair was winding up its run. Labor Day is seen as the unofficial end to summer in Onondaga County. The weekend is a time for travel, picnics and preparing to send the kids back to school. As Sunday night merged into Monday morning, a line of strong thunderstorms crossed into Oswego County from the northwest.[11] Moving at 40 to 60 miles an hour, the thunderstorms generated a derecho, a straight-line wind that generated winds from 70 to 115 miles per hour, the equivalent of an F-1 or F-2 tornado on the Fujita scale. Thousands of trees were uprooted, buildings were damaged and three people were killed. Two of the fatalities occurred at the New York State fairgrounds.

The forty-seven-minute storm cut power for 250,000 people in Upstate New York—157,000 of them resided in Onondaga County. Many of them were without electricity and phone service for over a week.[12] The storm was capricious in what it destroyed or didn't. Most of my neighbors had trees down, including one who had a tree stave in his roof. Our three Norway maples weathered the storm with only the loss of a few branches. Some streets were impassible for days. The City of Syracuse spent an estimated $23 million in cleanup, overtime and repair to damaged city buildings. Saint Lucy's Catholic Church on the city's west side lost its steeple. It was never replaced. President Clinton declared the area a disaster area. As bad as it was, it could have been so much worse!

Had the storm struck during daylight hours, while the state fair was crowded with fairgoers, the death toll could have been horrific. The storm struck at an hour when most people were home and not traveling. It cut a thirty-mile swath of destruction through the county and shattered illusions that Onondaga County was immune from the havoc wreaked by strong winds. We awoke to the sound of chainsaws cutting up massive tree branches on September 7, 1998. The Labor Day Storm is the storm my generation's children will tell their offspring about when tales of Mother Nature gone wild circulate over Thanksgiving dinner. The damage has long since been repaired, but every so often, if you know where to look, you can still see vestiges of the damage.

Three storms changed the landscape, shattered our illusions and made history in Onondaga County. No one is alive who witnessed the tornado of 1912. It shaped Long Branch Park. There are still people around who witnessed Hurricane Hazel and the Labor Day Storm. As the stories are told of these storms, they will perhaps grow with the telling. That is to be expected but perhaps a bit sad. The three storms were powerful enough to need no embellishment. Sadder still would they be if people forgot.

3

BURNING BRIGHT

Fire Courses through the County

As I type this, firefighters battle a lumber yard fire. Conflagration has shaped and reshaped Syracuse and the county. The fire in question has damaged several buildings and will likely result in millions of dollars in damage. From the dawn of time, fire has been used to benefit man. It has also caused damage and death. In Onondaga County's earliest days, fires were fought with buckets and whatever water was at hand. In the Salt City, that meant the Erie Canal and fifteen cisterns that the village had established for combatting fire. Syracuse did not have a fire department until 1834. In that year, the village had three volunteer fire companies and bought a hand pumper for the then princely sum of $925.[13] They didn't have a long wait to test their mettle.

On March 15, 1834, the B.F. Rogers store caught fire. The entire block and several wooden houses were consumed in the inferno. The firefighters did all they could, but the close quarters of the predominantly wooden buildings conspired to thwart their efforts. The Rogers fire was a training exercise for worse to come.

Fire is always terrifying, but perhaps it is no more so than when you are staying in a hotel. The Leland Hotel, touted by some as the finest hotel in Syracuse, housed 150 guests and several employees on October 16, 1890. As the guests slept, the clock struck 12:45 a.m. Lewis Leland, a hotel owner in his own right, was visiting his cousin Watson Leland, who owned the Leland Hotel. Lewis left the office to use the restroom, spotted the flames and sounded the alarm. Employee Clark Bridgeman notified the telephone

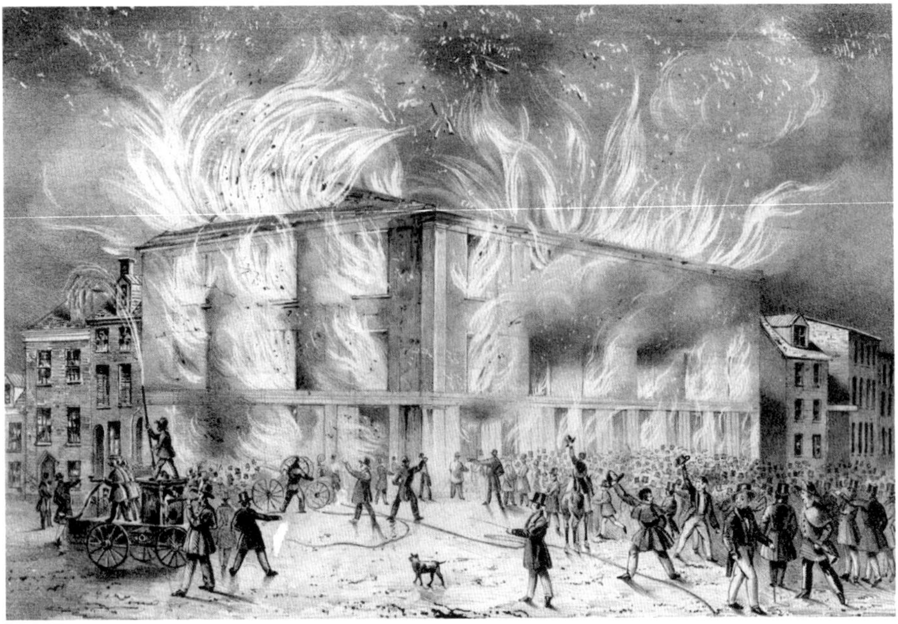

Destruction of Pennsylvania Hall showing a hand pumper similar to the one used at the gunpowder explosion. *Courtesy of the Library of Congress.*

office that the Leland was ablaze. John Bridgeman sounded an electric alarm that rang throughout the stricken hotel. Lewis and another employee went floor by floor rousing guests. Despite their efforts, at least 15 people died and several were injured in the conflagration, which destroyed the six-story building.[14] Many of the dead perished when they leapt out of the windows of the burning hotel. Eight days later, search teams were still poring over the ruins in search of more bodies. The Leland was worth $80,000, and its furnishings were valued at $150,000. In total, the damages were estimated at $250,000. These days, people remember the Collins Block fire, and the Leland Hotel fire is largely forgotten.

Many towns throughout Central New York and the United States boasted opera houses.[15] These venues provided entertainment to a population with increasing leisure time. In Syracuse, the place to see and be seen in the nineteenth century was the Wieting Opera House. The Wieting Opera House did indeed host operas, but it also presented lectures, church services and political conventions. The Wieting went through three incarnations in its reign as Syracuse's premier entertainment venue. First built in 1852, the then Wieting Hall was gutted by fire in 1856. Dr. John Wieting, the owner of the building, rebuilt and expanded the venue. It reopened in 1870 but was consumed in an inferno on July 19, 1881. Again, Wieting rebuilt. Fire immolated the building in 1896. Wieting's widow rebuilt, and the new Wieting Opera House opened its doors in 1897. The Wieting soldiered on for another thirty-three years, until it was demolished in 1930. The spot is now occupied by the Atrium.

There are years when it seems like the fates conspire against mere mortals. The year 1907 seemed to be just such a year for Syracuse and Onondaga County when it came to fires. Early one Sunday morning, February 10, 1907, the Mowery Hotel caught fire. The blaze started when an explosion occurred in the hotel's saloon, and it spread quickly. The hotel was originally intended to be an apartment building but quickly converted to a hotel for travelers. The building was a total loss. Initial estimates of damage were $250,000. There were no deaths in the fire and very few injuries. William Murphy, the hotel's elevator operator, saved several people when he kept the elevator running as long as he could during the blaze. You won't find Murphy's name on a plaque, but he is one of Syracuse's unsung heroes.

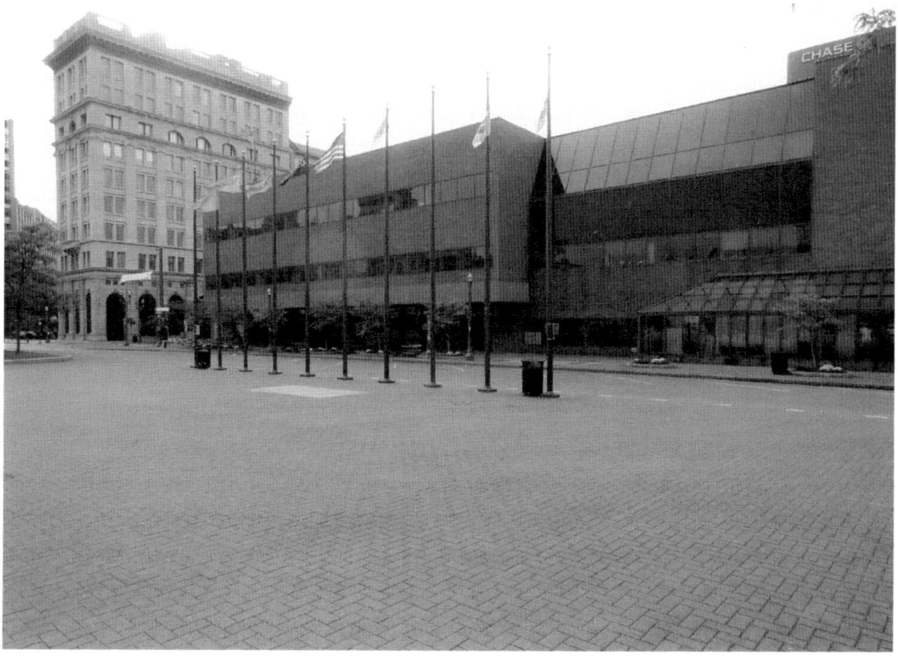

The Atrium now stands where the Wieting Opera House once stood. Fire damaged the venerable opera house several times before it fell to the wrecking ball in 1930. *Courtesy of Neil K. MacMillan.*

As winter approached Syracuse in 1907, another fire shook the city and county to its core. The blaze could have been a tragedy on par with the worst in Onondaga County. The House of Providence Orphanage in the town of Geddes was run by the Sisters of Charity, and by all accounts, the good sisters were highly capable in caring for their charges. On the morning of Sunday, December 8, 1907, 152 orphaned boys were gathered for breakfast. A housekeeper told one of the sisters that she smelled smoke. The boys were gathered in a playroom, and the sisters began an investigation. A wall of the orphanage collapsed. The sisters marched their charges out of the burning building and kept the boys from returning. Providence House was rapidly engulfed. The Syracuse Fire Department responded quickly to the blaze, but their efforts were hampered by a lack of hydrants or a suitable water supply. The boys escaped with the clothes on their backs. That none of the orphans was injured in the fire was attributed to the hour and the calm courage of the nuns charged with caring for them. The three-story building was a total loss.

The Bastable Block also housed a venue to rival the Wieting. The original building and the second Bastable building were both the brain children of impresario Stephen Bastable. The Bastable anchored Hanover Square, standing where the State Tower Building now stands. It was a structure famous in its day for the entertainments offered and the three fires that plagued it. In addition to serving as an entertainment venue, the Bastable also functioned as an office building. The first fire struck the Bastable Block in 1862. At the time, the building housed shops and offices. Bastable's loss was assessed at $40,000, and merchandise valuing $75,000 was lost. There was not enough insurance to cover either, but Bastable rebuilt.[16] When Stephen Bastable opened the rebuilt Bastable Block, it included a theater, as had the first building. The venue featured many of the preeminent acts of the late nineteenth and early twentieth centuries. On November 20, 1891, fire gutted the Syracuse landmark. Although valiant, firefighters' efforts proved to be in vain, and the building was destroyed. Controversy was added to the loss of the building when one of Syracuse's fire commissioners called for hearings to determine if fire companies had been effective in their use of extant equipment. The controversy started over the request of the department to buy a water tower to help fight multistory structure fires. Commissioner Kanaly withdrew his objection to the water tower and then called for the investigation into the conduct of firefighters. The Bastable family rebuilt yet again.

The new Bastable Block served for thirty-two years when fire struck the venerable edifice for the final time on February 12, 1923. Within a half hour, the fire claimed three lives and injured several others and left the Bastable a smoldering ruin. In the subsequent investigation into the fire, residents of the building claimed they did not know where the fire escapes were. While most witnesses praised the efforts of the fire department, a few were critical. One victim died when he jumped from a window and missed the net rescuers held to catch him. Demolishing the remains of the Bastable Block proved problematic as well. Twelve attempts were made to dynamite the ruins, with limited success, and resulted in damage to nearby buildings. The remaining walls were finally pulled down with ropes and brawn. The heyday of opera houses had passed. The Bastable family sold the land to a developer, and four years later, the Art Deco State Tower Building opened for business.

Sixteen years later, fire would visit the Collins Block, which was located next to the State Tower Building site of the fatal 1923 Bastable Block fire. No fire since the 1841 gunpowder explosion was as devastating. In the early

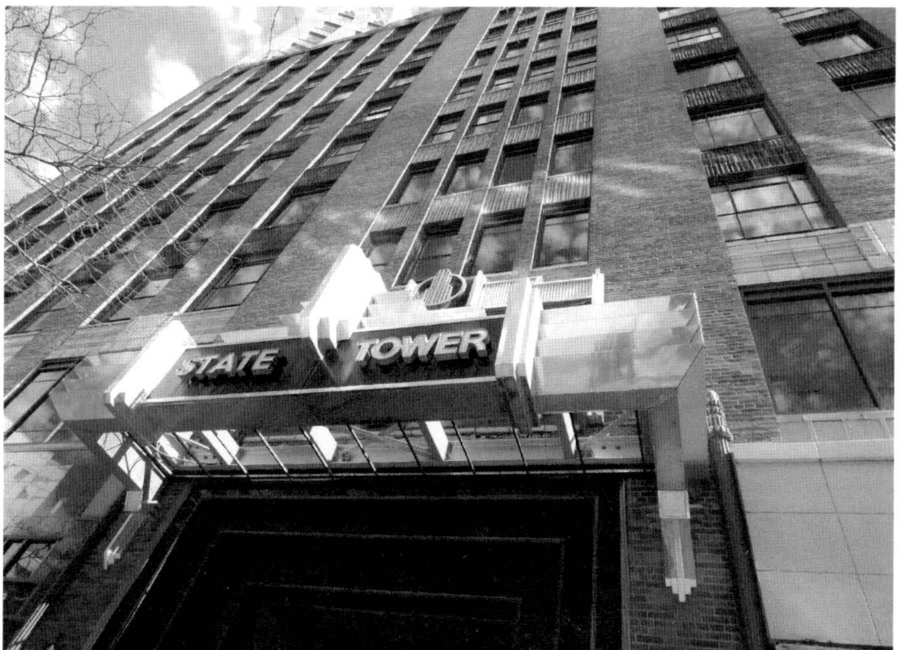

The State Tower Building was erected in 1927 on the site of the Bastable Block, which burned in 1924. *Neil K. MacMillan.*

morning hours of February 3, 1939, the Collins Block at 231 East Genesee Street caught fire. The fire department made headway in fighting the fire, although it had substantially damaged the building. The fire was out and the firefighters were leaving when the building collapsed. Eight men were buried under the falling building. As in the great gunpowder explosion, speculation was rife as to the origin of the blaze. Unlike Syracuse's first major fire, investigators had the tools to perform a comprehensive investigation. Almost immediately, rumor focused on striking waiters at Ada Keep's restaurant, which was housed in the Collins Block.[17] Just as quickly—though organized labor was still controversial in the late 1930s—the rumors were discounted. Investigation revealed that faulty heating equipment was the culprit.

Firefighters John Frost and Leo Dixon were cited for heroism in their efforts to rescue their fallen comrades. Frost rescued fellow firefighter William Moore, while Dixon was cited for pulling his brother M. Gregory "Mike" Dixon from the wreckage of the Collins Block. Unfortunately, Mike Dixon was among the dead.

Three other buildings—the Flaherty, Moose and Callahan buildings—were closed after being deemed unsafe. Firemen were on twenty-four-hour duty during the funerals of their fallen friends. It took two days for the first body to be recovered, that of Lieutenant Albert G. Young. Amid the toil of digging through the rubble, a former psychiatric patient confessed to starting the fire. It was quickly determined the young man was mentally unstable but not responsible for the inferno. In the aftermath of the Collins Block tragedy, several spectators were arrested for throwing snowballs at rescue crews. The fire brought change as well. In 1939, firefighters received very little training compared to the firefighters of today. Among the things learned today is a two-week session on building construction so firefighters can identify structural problems in an effort to avoid a repeat of the 1939 disaster. New trainees are taught to follow procedure and be aware and vigilant of their surroundings. The City of Syracuse established Firefighter's Memorial Park on East Fayette Street to honor the Collins Block dead and all firefighters who have fallen in the line of duty. A plaque at the site also commemorates the fire and the fallen heroes of the Collins Block fire. As we inch closer to the eightieth anniversary of the fire, there will be more fires. Hopefully, none will so devastate the fire department or the community.

An overnight fire injured two maintenance men and destroyed the K-Mart in Mattydale on April 13, 1970. Damage was estimated to be between \$3 and \$10 million. Eight fire companies battled the blaze. The heat of the fire was so intense it twisted some of the building's steel girders

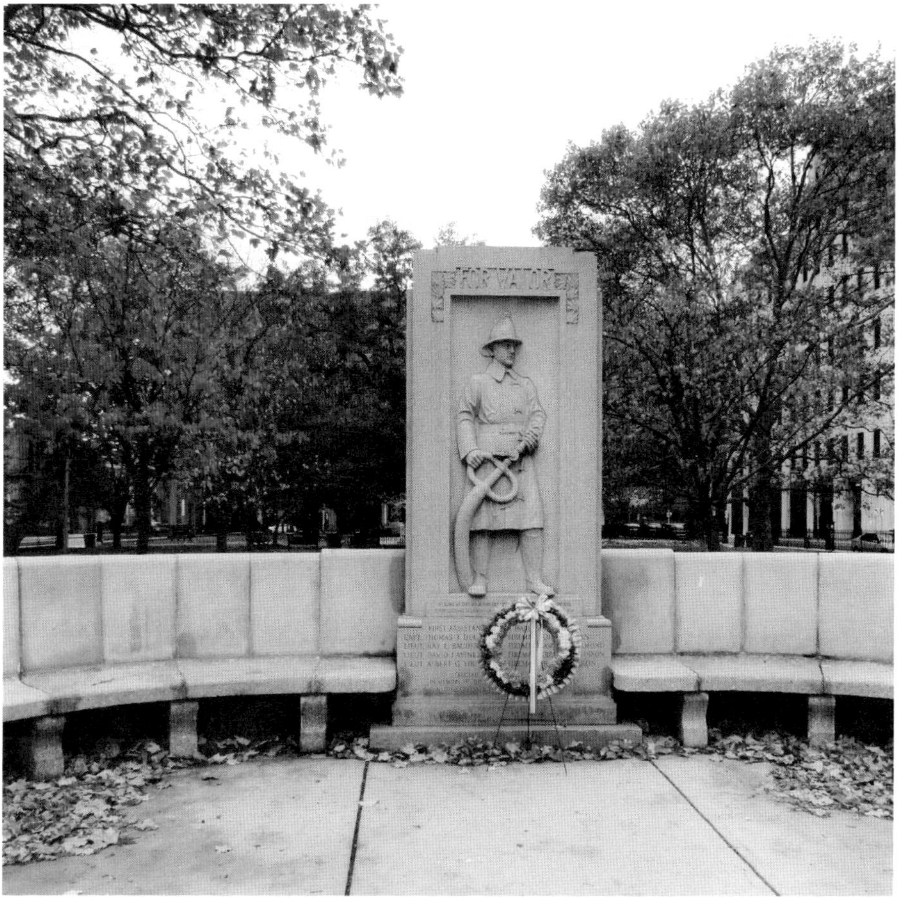

The Fayette Park monument to Syracuse's fallen firefighters. The monument was erected to honor the eight firemen killed in the Collins Block fire in February 1939. *Neil K. MacMillan.*

and cooked off ammunition in the sporting goods department. Most of the roof and two walls of the store collapsed. The blaze started in the appliance section and spread rapidly due to the open nature of the store. The retail giant rebuilt the store—a total loss with the exception of several garbage cans and eleven small rowboats that had been stored outside. K-Mart is still a focal point in Mattydale, and the plaza it shares with several other businesses is known by locals as "K-Mart Plaza."

Fire has been a boon to Onondaga County and sometimes a curse. As long as we build edifices of flammable materials, it will continue to be so. We can only hope that we'll see no more losses as we did in the Leland and Collins Block fires.

4
LET LOOSE THE DOGS OF WAR

Goose Van Schaik and Pan-Am Flight 103

What is now Onondaga County was once the seat of the Haudenosaunee Confederacy. At the start of the American Revolution, the six-nation confederacy strove to maintain neutrality in what it saw as the white man's conflict. The history of the Onondaga and all Native Americans is one of mistrust and broken promises. Neither culture understood the other when they encountered each other, starting with European colonization. In previous conflicts, the entire Haudenosaunee alliance cast its lot with the English against the French in their struggles to dominate North America. The alliance with the English was more of a shotgun wedding than a mutually beneficial friendship. Decades of sporadic warfare against their traditional enemies, the Algonquin—who were supported by the French—impelled the reluctant nations of the Haudenosaunee to side with the English. When war broke out between King George the Third and his rebellious colonies, the nations of the Haudenosaunee Confederacy were in a quandary. All six nations wanted to maintain their neutrality. They saw no benefit in choosing sides in what they saw as a family conflict.

Both British and American agents tried to bring the six nations into their camp. While the Haudenosaunee tried to stay neutral, factions throughout the alliance lined up with opposing sides. Eight Onondaga chiefs sided with the Oneida and Tuscarora, who generally supported the rebellious colonies. They argued that the council fire was extinguished and advised choosing sides. The Onondaga Nation rejected that advice and countered that no decision to choose sides or extinguish the council fire could be made without all chiefs being present. By January 1779, when the eight Onondaga chiefs gave their advice, neutrality was no longer an option for the Haudenosaunee Confederacy.

From the start of the American Revolution, Loyalists and their Native American allies raided frontier settlements. Quarter was not given, and what were considered rules of "civilized" warfare were not adhered to. The Wyoming Valley Massacre and those of Cobleskill and Cherry Valley, New York, enraged the Patriots. Determined to end the depredations, General George Washington ordered Major General John Sullivan and Brigadier General James Clinton to "take the war to the enemy to break their morale." The tactics were to be the same used by the Loyalist and Native American raiders. The task of suppressing the Onondaga was given to Colonel Goose Van Schaik.

On April 16, 1779, Van Schaik assembled his troops and gathered supplies at Fort Schuyler, which is now Fort Stanwix, in Rome, New York. On April 18, 558 soldiers left the fort. The expedition lasted twelve days. Van Schaik and his men cut a swath through Central New York like a scythe through wheat. Among the native villages destroyed was the main Onondaga village of Onondaga Castle, located in what is now the Valley District of Syracuse.[18] Stores were destroyed, crops and villages burned and noncombatants taken prisoner. The raid broke the back of Haudenosaunee resistance to Washington's forces. Goose Van Schaik's raid was seen as the most effective of the Sullivan/Clinton expedition.

When the Revolution ended, the British abandoned their Native American allies. Several treaties eroded the position of the Onondaga and the Six Nations. When war came between the young United States and the British in 1812, the Six Nations stayed neutral. A friend of mine who is Mohawk reminded me that the Haudenosaunee never surrendered to the United States. That is absolutely correct, and if the British had not thrown their allies under the bus, the Haudenosaunee might have continued their resistance to the United States or at least been able to negotiate better treaties for themselves. The Onondaga were driven off most of their land and marginalized by the Americans. In 1795, Ephraim Webster was given a square mile of land by the Onondaga, and American settlement of Onondaga County began.[19] War breeds a number of outcomes, and what is seen as victory for one side is disaster for the other. Goose Van Schaik's raid was a disaster for the Onondaga, but the Onondaga Nation remains a vital, vibrant part of Onondaga County and a sterling example of courage, tenacity and endurance in the face of adversity. The raid was the last time

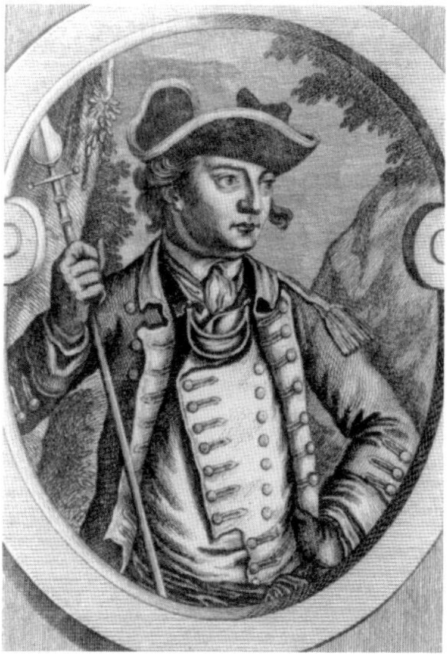

Major General John Sullivan led the expedition against the Haudenosaunee during the American Revolution. His subordinate, Colonel Goose Van Schaik, led the thrust against the Onondaga in the spring of 1779. *Courtesy of the Library of Congress.*

combat troops fought on the county's soil. It would not be the last time warfare touched our lives.

There was a time before September 11, 2001, when we in Onondaga County felt relatively immune from the ravages of terrorism. We had a steady faith that we were too far away and too well protected to suffer a terror attack.[20] December 21, 1988, promised to kick off the Christmas holidays, as in previous years, with a travel boom.

Pan-American Flight 103 left London full with people returning home for the holidays. The majority of the passengers were Americans. Among the travelers were thirty-eight students from Syracuse University, several students from other Central New York Colleges, including Cornell University and SUNY Oswego, and a couple from Clay, New York, who had attended a wedding in London. The students were enrolled in a popular overseas study program. The Boeing jumbo jet lifted off for the long flight to New York carrying eager passengers westward.

Pan-Am 103 was a marked flight. Islamic radicals had threatened to bring down a United States–bound Pan-Am flight from Frankfurt, Germany.[21] Incensed by American support for Israel and U.S. military attacks against terrorists, the radicals targeted an aircraft they knew would carry many Americans. Someone placed a bomb aboard the flight. British authorities believed the bomb, which was disguised as a tape recorder, had been placed on the flight in Frankfurt.[22] The West German government stoutly denied the British claim.

As the aircraft passed over Lockerbie, Scotland, the bomb exploded, and 270 were killed. Of those, 42 were from Central New York or attended universities in the area. It was the worst air disaster in British aviation history.

Only one person was convicted for the bombing. In January 2001, Abdelbaset Ali Mohmed al-Megrahi was convicted by Scottish authorities of murder, conspiracy to murder and breach of the aviation security act of

1982. He was sentenced to life imprisonment, and his conviction was upheld after several appeals. The Libyan national was released on compassionate grounds in August 2009 after being diagnosed with terminal prostate cancer. He died in 2012.

The conspiracy theories started before the smoke cleared from the Scottish countryside. Did al-Megrahi bomb Pan-Am 103, or were others responsible? Abdelbaset al-Megrahi was convicted under Scottish law, and his appeals were consistently denied.[23] While many in Libya claimed the verdict was unjust, it is likely Scottish investigators got their man. Investigators initially thought a splinter group of the Palestine Liberation Organization (PLO) or the Iranians might be responsible. Some claimed that Israel's Mossad planted the bomb to further implicate its enemies. One theory even claimed that South Africa's apartheid government was responsible. While there may have been inconsistencies in the conviction of al-Megrahi, it is notable that the Ghaddafi regime paid $2.7 billion in compensation for the bombing and other acts of terrorism.

We live in a world where security checkpoints and the threat of acts like the bombing of Pan-Am 103 are an everyday part of life. The bombing pales in the wake of the attacks of September 11, 2001. Pan American airlines declared bankruptcy in 1991 and divested itself of all holdings. The airline was yet another victim of the bombing and the subsequent lawsuits and government investigations. Airline security was lax in the days before the bombing. The government did not publicize the warnings that preceded the bombing, an action that New York Republican senator Alphonse D'Amato strongly condemned on the floor of the Senate. There are memorials to the victims in Lockerbie, Arlington National Cemetery and at Syracuse University. The university and the community remember the victims each year on the anniversary of the bombing.

Two wars brought devastation to the area. In the case of the Van Schaik expedition, the war was declared and the enemy well known. The carnage brought about crippled the ability of the Six Nations to wage effective war. In the case of Pan-Am 103, the war is an ongoing one fought in the shadows by people who are all too often nameless faces. The devastating loss still haunts Onondaga County. Both provide lessons for us in the twenty-first century.

5

IF YOU BUILD IT, WILL IT STAND?

Building Collapses in Onondaga County

In the movie *Field of Dreams*, Kevin Costner's character is told by an ethereal voice, "If you build it, they will come." The voice was referring to the great baseball players of the golden age of baseball. All too often, when we erect structures, gravity decides to intervene. Sometimes, these collapses are due to natural phenomena. Here in Onondaga County, usually it is due to human error. In fact, it is commonly due to shortcuts taken during the construction of the ruined edifice. Let's face it: the unscrupulous contractor has been with us since we started putting up buildings. Money is a motivator in these instances. In an effort to get the job, the miscreant will underbid and, to gain the maximum amount of profit, will use inferior material and unacceptable shortcuts. Once in a while, the chickens come home to roost quickly. All too often, however, it takes years before the bad construction comes to light. Sometimes, it boils down to the fact that some people just should not be allowed to play with tools.

In 1872, the congregation of the First Baptist Church dedicated a new church building. The Gothic structure was located at the corners of Jefferson and Montgomery Streets. As the church was designed by prominent architect H.N. White, the congregation expected that the new edifice would see them through a lifetime of worship. It is said to err is human, to forgive is divine. While under construction, the lower chords of the roof joists for the new church were spliced upside down. In addition, the joists had been made with hemlock lumber instead of the usual and called-for pine. Had the architect's design been scrupulously followed, even this

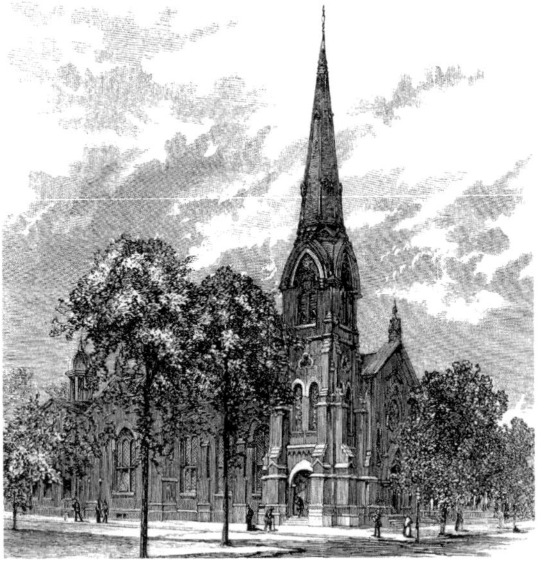

The Central Baptist Church was a fixture on what is now Columbus Circle in 1874. *Courtesy of the Library of Congress.*

might have not been a problem. The original design called for the second-story floor to be supported by pillars. At some juncture, the pillars were done away with, and the support for the second story floor was provided by rods attached to the joists. H.N. White protested this decision vigorously, but he was overridden. He refused to continue on the project, but the completed building was dedicated with prayers and celebratory hymns of thanksgiving.

On the evening of June 23, 1874, the congregation of the Central Baptist Church held a children's concert and strawberry festival. The event was well attended by the congregation and others from the community. The festivities were in high gear by 9:20 p.m. Then the floor gave way. In a moment, the festive atmosphere became tragically chaotic. Fourteen lives were snuffed out in that instant, and one hundred were injured.[24] Two children were presumed dead. The two missing children, sisters Kate and Nellie Moseley were found hours later. They had been taken home by their Sunday school teacher. The majority of the people attending the festival were on the collapsed floor. Almost as rapidly as the floor collapsed and searchers began combing the ruins, buildings in downtown Syracuse were draped in black mourning crêpe. The tragedy was news throughout the nation. *Frank Leslie's Illustrated Newspaper* carried a drawing of the wrecked church interior on its cover for the July 11, 1874 edition.

Both the architect and the building committee were criticized for "acts of omission." Looking back with hindsight, we can second-guess the investigator or the grand jury. White was censured because his account of what transpired with the floor supports was disputed by William Dickenson and Manning Palmer. The builders received the severest censure for failure to meet requirements for the building's safety. James Crow sued the church over his wife's death. The stresses put on the joists and supporting rods were more than the structure could withstand, and the hemlock used in constructing the joists was riddled with dry rot that had not been detected when the joists were constructed. The pillars that the building committee dispensed with might have forestalled the tragedy. It is one of the questions we may never have an answer to. At the memorial service, the choir and congregation sang "Nearer My God to Thee." Thirty years later, the hymn would gain fame as the song played by the ship's band as the RMS *Titanic* sank.[25]

The Central Baptist Church floundered on for some years but eventually sold the property. The First Baptist Church later built a church on the site. The now iconic Mizpah Tower is a familiar site in Columbus Circle. It is now empty, a victim of a more secular outlook. It is directly across Jefferson Street from the Cathedral of Saint Mary's, built in the year the Central Baptist Church collapsed. The collapse was one of the worst disasters in the county's history, but it is little known today.

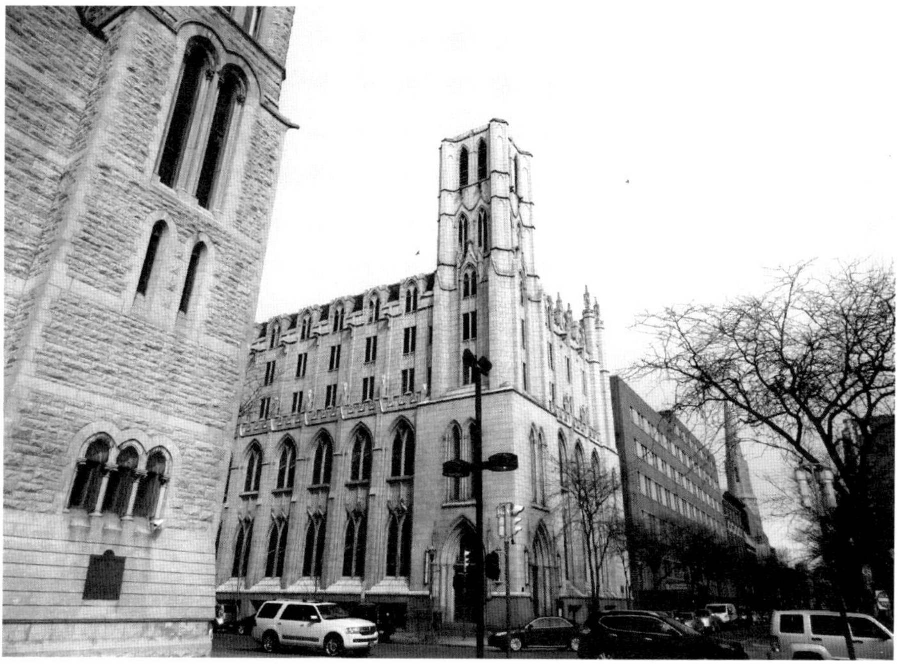

The First Baptist Church is also known as the Mizpah Building and is used by the YMCA. It stands on the same site as the Central Baptist Church. *Neil K. MacMillan.*

Syracuse once boasted forty breweries that put out copious amounts of quality beer. Names like Bartel, Haberle, Ryan and Greenway were well known and enjoyed throughout Onondaga County. Indeed, Greenway brewery could boast that its product was consumed throughout the world—as far away as Australia and China. John Greenway started his brewery in 1858 and ensconced it in a building along the Erie Canal between Water and Canal Streets. Greenway Brewery was a generous employer, going so far as to build apartments for its employees on Hawley Avenue on Syracuse's near north side. The brewery hosted an enormous barbecue for the city in 1870.

On January 5, 1880, the fifth floor of a section of the huge Greenway brewery collapsed, carrying the floors beneath with it. Two workers suffered minor injuries. The brewery lost 700 bushels of corn and wheat, 175 bales of hops, 5,000 bushels of Canada malt and 2,000 bushels of powdered malt that were pitched into Onondaga Creek and swept away. John Greenway repaired the damaged structure. Losses were estimated at $20,000.[26]

In spite of the brewery's prosperity, there was more adversity in store for Greenway's. On July 30, 1907, a keystone for the Erie Canal viaduct fell into Onondaga Creek, draining the canal.[27] The rushing water weakened the foundations for the Amos Mill building and the Greenway brewery. The walls adjacent to the stricken canal collapsed the next day. Again the building was repaired.

On September 6, 1913, the Greenway brewery caught fire and burned to the ground. The company moved to Hiawatha Boulevard. With the advent of Prohibition in 1920, all the area breweries suffered financial loss. Some never reopened. Greenway did and soldiered on through World War II before closing in 1952. Its competitor, Haberle, lasted to 1964. The golden age of brewing in Syracuse survived Prohibition but could not compete with the mass production of breweries like Miller and Budweiser. Miller no longer brews beer in Central New York. The recent years have seen the rise of craft breweries, and Syracuse boasts a few craft breweries, including Empire Brewing, Clark's Ale House and Middle Ages brewery. They may not send beer to China like Greenway did, but I believe John Greenway would approve nonetheless.

There will continue to be building collapses due to natural causes and people's greed and incompetence. If we learn nothing else from the Central Baptist Church and Greenway brewery collapses and we heed that lesson, Onondaga County will be better off. If we do not, more people will die and more damage ensue.

6

RIDING THE RAILS

Train Wrecks throughout Our History

Historians will tell you the Civil War changed how we view America, and they're right. In the antebellum period, we referred to our country in the plural—"The United States are…" Since the end of the Civil War, we refer to our country in the singular. It seems like a small thing, yet it was a radical shift in our thinking. So, too, the railroads changed America and Onondaga County in ways that still have ripple effects to this day. Travel by train allowed the vast expanses of the Great Plains to be opened up for settlement. Here in Onondaga County and in New York State, the railroads supplemented and then replaced the Erie Canal as the major means for transporting goods and people. The day of the railroad as the premier means of transport passed as well. The trains still run, but there aren't as many now.

Like any vehicle, trains have been beset by safety problems. The history of railroads is littered with the wreckage of train cars. The history of rail travel is equally littered with train wrecks and accidents. To this day, people try to beat the train. It rarely works. The train doesn't stop on a dime. People walk along railroad tracks or play on them. Tons of metal rolling at even moderate speeds turn flesh and bone to pulp. Trains get derailed. Usually, this is because of excessive speed or faulty rails. In one instance, it was due to an overpass collapse.

Among the hazards facing trainmen 157 years ago were livestock. These days, we have lights, bells and barriers to keep vehicles from becoming wreckage. Farmers have barbed wire to keep their cows in their fields. On

May 18, 1859, at the Cold Spring crossing near Jordan, New York, a cow jumped into an oncoming train between the tender and the baggage car. The baggage car and two coaches were derailed. The conductor, Thomas Gifford, was killed, and fifteen others were injured. One of the injured, Samuel Slossen of Ohio, was thought to be mortally wounded, but he subsequently recovered. A disturbing lesson was being learned in the nascent days of rail travel.

Six years later, another coach went off the tracks just three miles from Syracuse. Fifteen were injured in the derailment, including Captain Charles Lewis of the Sixteenth United States Infantry and prominent Liverpool salt baron Thomas Gale.

Fifteen years later, Engine No. 115, hauling about fifty-five cars, drove into a banked underpass filled with approximately three feet of water. The engineer thought the train would make it through the minor flood, and it did. The problem arose when the boiler fire was put out. Though there were no longer flames, some of the water vaporized as steam and blew the boiler door open, injuring four men: engineer Joseph Stoutenger, fireman Engene Sanford, conductor John Mangan and a passenger, John Groff.[28] Groff suffered substantial burns and later died of his injuries. The train managed to make it out of the underpass in spite of the boiler fire being out. Sanford suffered only minor burns, but the others were seriously injured.

In 1847, Stephen Hayes from Taylor, New York, in Cortland County and Phillip Simon, a salt boiler from Liverpool, New York, were killed when the horse-drawn lumber wagon they were driving was struck by a train locomotive and tender at the DeWitt Plank Road crossing. The DeWitt Plank Road crossed the tracks at an extreme angle, and witnesses claimed that trains could not be seen at a distance of six rods, or ninety-nine feet.[29]

Can overpasses be notorious? It sounds like a facetious question, but it really isn't. The Park Street overpass was the site of two derailments in as many years. A New York Central freight train, Extra 1087 East, was pulling 106 cars and a caboose at thirty-eight miles per hour. At impact, 23 cars left the tracks, and 6 of the derailed cars demolished the Park Street overpass. The derailment was caused when the rear right journal of the 45th car failed. A signalman at Signal Station 2 noticed the train had a hot journal—later to be determined as the one that failed—and passed that information to Signal Station JG. The lever man at Station JG tried to warn the train crew as the accident occurred. Signalman Salem LaHood was killed when the signal station was demolished by the train. He was the only fatality and the only injury. The failed journal had been rebuilt in January 1957 and installed the morning of the crash. It had no history of overheating before. LaHood's widow was awarded $37,500 by a federal court.

On November 17, 1959, history repeated itself when a New York Central freight train derailed four hundred yards from the Park Street overpass. Fortunately, no one was injured in that derailment.

Lyncourt is a fairly small suburb on Syracuse's northern fringes. It isn't a place thought of when we think of catastrophe. On February 5, 2001, that all changed. Two tracks pass through the town of Lyncourt. One of these was closed for maintenance. A CSX freight train was traveling at seven miles per hour. Approaching on the same track was an Amtrak train filled with passengers nearing Lyncourt at fifty-eight miles per hour. Signalers warned the train that the speed limit for that section of track was fifteen miles per hour and to be ready to stop, as there was another train ahead on the same track. The Amtrak train struck the CSX freight at thirty-eight miles per hour. Sixty-two passengers and crew on the Amtrak train were injured when the two trains collided. Fortunately, no one was killed. Almost as quickly as the accident occurred, some unscrupulous lawyers began handing business cards to victims, a severe breech of ethics and reason enough to be disbarred. The lawyers who eventually filed a class-action suit on behalf of the victims were not involved in those shenanigans and spoke out vigorously against those who were. The class-action suit asked for $75,000 for each victim. The NTSB found the cause of the accident was engineer error. The track under maintenance opened seventeen minutes after the collision, and the crash caused $280,000 in damage. Had the engineer been following procedures for the actual signal displayed calling for him to slow to fifteen miles per hour or even the signal he thought he saw calling to restrict speed to thirty miles per hour, the accident would not have occurred. Inattention to his job caused the wreck, and sixty-two people paid for it.

There is a reason the crew members on trains, planes and buses remind us to remain seated until the conveyance has come to a full and complete stop. Sometimes we don't listen. Impatience gets the best of us, and we think we know better than the crew members. Usually, the only thing that suffers is our dignity. In the village of Manlius, a German man paid the ultimate price for his impatience. Warned by his fellow passengers that it was not safe to exit the train, he ignored their entreaties. The train was still backing up as he alighted from the coach. He slipped and the train ran over him. He was killed instantly.

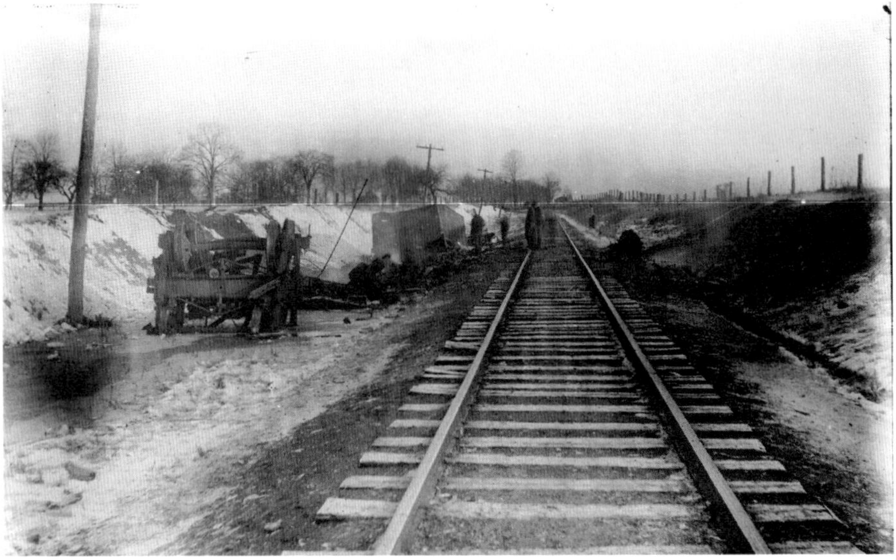

A train derailment in the early twentieth century. Derailments continue to this day and are usually the result of excessive speed. *Courtesy of the Library of Congress.*

On August 17, 1918, little Viola Stark was going to the store. Intent on her errand, she didn't see the train approaching the Walnut Avenue crossing. A neighbor called out a warning. Viola hesitated, and the train struck and killed her. It is possible that had she not been warned or not hesitated on the tracks when warned that she might have lived.

Ray Fortner should have known better. While cutting through the East Syracuse railroad yards, the teenager and some friends darted between two train cars. Fortner's friends thought he had made it through the train and took off without telling them. In fact, he had been cut in half by the train. His body was found by Conrail employees. The case was noted in the newspapers because there seemed to be a jurisdictional dispute between the East Syracuse police and the Conrail Railroad police.

There are still unmarked railroad crossings in the United States. I find that amazing. Forty-six years ago, a family was ravaged by the dangers of an unmarked railroad crossing. The St. Hilaire family was returning from a Sunday visit and not far from their home when they crossed the railroad tracks on Saintsville Road in Minoa. Their car was struck by a train going fifty-eight miles per hour in a zone rated at eighty miles per hour. Three members of the family—Robert, aged thirty-five; his wife, Keitha, aged thirty-four; and one-year-old daughter Susan—were killed instantly. The couple's other two daughters, fifteen-year-old Patricia and eleven-year-old Robin, were seriously injured. Robin died nineteen months later. She remained in a coma from the time she was injured. Tragedy visits in many forms. It visited in its most horrific with the St. Hilaire family.

History teaches us many lessons. When it comes to trains, perhaps the most vivid lessons are those provided when things go horribly wrong. The accidents I mention in this chapter can be teaching tools for all of us. Indeed, if they are not, then the people injured and killed in them have died in vain.

7
WATER, WATER EVERYWHERE

Floods

There is such a thing as too much of a good thing. Here in Onondaga County, we don't generally see the massive flooding that plagues other parts of the country. That doesn't mean the county is immune to flooding; we just don't see it on the scale that some places do. We generally get ten feet of snow per year, and when it melts, it has to go somewhere. Usually, we have brief thaws and don't get all of the season's snow in one fell swoop. We are affected by tropical storms and bad weather as other places are, but usually, it is not severe. The threat, however, is always present.

Flooding in late March 1865 disrupted canal and rail traffic throughout Central New York. Water on the Seneca River rose to levels never seen before. Baldwinsville residents with boats were charging $3, approximately $210 today, to ferry people across the flooded areas. Closer to Syracuse, rail traffic was halted due to tracks being under water. The *Syracuse Daily Courier and Union* of March 24, 1865, called the flooding the rise of "seasonal waters."

Rainy weather in September 1890 led to widespread flooding throughout Central New York. The new tiled drainage system installed at the New York State fairgrounds kept the venue from being inundated. Flooding in September is not usual in Onondaga County, and it wasn't 126 years ago. The flooding—due to prolonged rain—didn't keep the fair from opening on time.

Flash flooding is always a possibility when rains get heavy. One downpour struck the county in July 2015. One man died when he attempted to save a

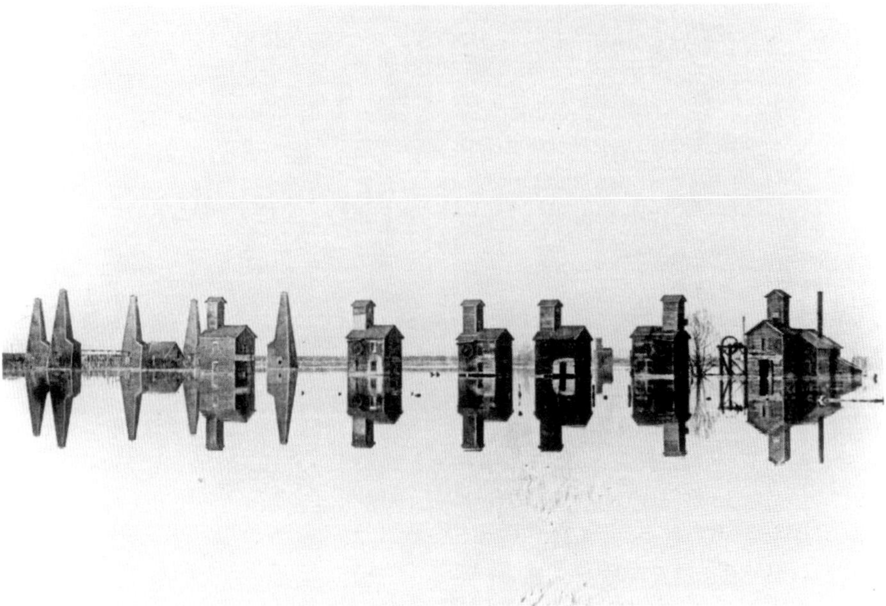

Flooded well houses along Onondaga Lake. Rising water has been a concern for the county throughout its history. *Courtesy of the Liverpool Public Library, from the Schuelke Collection.*

stranded motorist and he was pulled into an open manhole by the rushing flood waters. The sewage system in the city of Syracuse is designed so the manhole covers will blow off their holes to relieve pressure on the system and keep sewer pipes and, consequently, roads from collapsing.

Onondaga Creek was a lifeline for the Onondaga Nation in the days before European settlement of Onondaga County. The creek meandered from Cardiff Valley, through what is now the Onondaga Nation northward through Onondaga Valley and to Onondaga Lake. The creek cuts a watery path through the city of Syracuse. Historically the creek flooded every year as the snow on the hills surrounding the creek melted. Onondaga Creek is situated between steep hills. The valley was carved out by glacial ice during one of the ice ages that carved out deep ravines and valleys throughout the northern hemisphere. If the creek provided fish and a reliable highway, it could also become treacherous when the snows melted.

The flooding of March 1865 mentioned earlier is a point in case. Buildings along the creek were inundated, and two sportsmen intent on taking advantage of the flooding to shoot muskrats driven from their dens along Onondaga Creek were almost killed by those same waters. The intrepid duo ensconced themselves in an outhouse to use it as a blind to shoot the muskrats from. The rising waters carried their perch away, but one of the hunters escaped as the "necessary" left its footing. The other stalwart escaped when the outhouse smashed into a tree. Rail and canal traffic was disrupted throughout the county.

In March 1902, Onondaga Creek left its banks yet again. Although several basements along the waterway were flooded, there was no serious damage. In October of that same year, residents living along the creek brought suit in the state supreme court against the City of Syracuse for damages done by the previous winter's flooding. The case was postponed. By this time, the city and county were looking for solutions to the annual flooding problem posed by Onondaga Creek. Solutions would take time.

Editorials in March 1901 lambasted city officials for failure to act on solutions to the yearly flooding. The complaints and the search for mitigation would occupy a good portion of the twentieth century. In July 1904, claims against the city reached $13,000. Onondaga Creek roared out of its banks again the last day of February 1910. This time, it wasn't just basements that

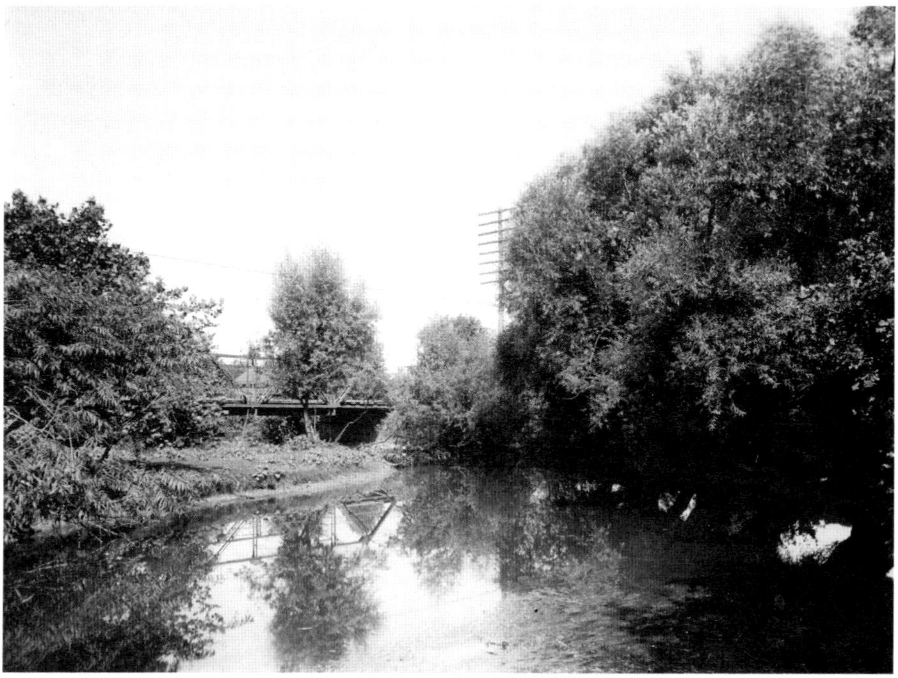

Onondaga Creek before flood control efforts changed its course. *Courtesy of the Library of Congress.*

suffered the intrusion of floodwaters. The towns of Onondaga Valley and Elmwood, along with the city of Syracuse, were submerged. Mrs. Alonzo Eigabroadt, an invalid, was carried from her home by rescuers. Damages ran into the thousands of dollars.

In July that year, a water main break sent a roaring plume of water fifteen feet into the air and turned South Salina Street into a lake. The torrent lifted a trolley off the tracks. Railroad and trolley traffic was severely disrupted, and lift bridges over the canal were not functional due to the drop in water pressure caused by the break.

Torrential downpours in July 1915 caused over $10,000 in damage and wreaked havoc on farmers' crops. The commissioner of public works vowed to honor all legitimate damage claims and that work would continue on Onondaga Creek. As late as 1930, engineers were warning that Syracuse's sense of security from a great flood was based solely on fantasy. The city, county and state worked to alleviate the flooding damage from Onondaga Creek. A dam was constructed on the Onondaga Nation's land, and the path of the creek was straightened and altered to minimize the damage wrought by spring floods. In addition, fencing was installed to combat the accidental drownings that occurred with increasing frequency in the creek. Flooding continued into the 1960s. A young boy drowned in the floodwaters of the creek in 1958. His body was not recovered for several days. Families living in Nedrow, just south of Syracuse, were evacuated in February 1961 as the creek surged over its banks again. In 1962, ground broke to widen, deepen and straighten Onondaga Creek in a continuing effort to control flooding. By February 1963, the Army Corps of Engineers had reported that the threat to Nedrow from flooding was reduced. By the 1970s, Onondaga Creek's makeover was complete and the flooding alleviated. I will not claim that the creek will never flood again. Mother Nature has a wicked, capricious sense of humor and may well send massive rain or snow again. For now, the creek stays in its banks.

Onondaga County doesn't usually get tropical storms. Normally, by the time the storms that ravage the coasts of Florida and the Gulf of Mexico get here, they have lost most of their punch and we get a few rainy days. Occasionally, we get a storm like Agnes. Tropical Storm Agnes formed on June 14, 1972, over the Yucatan Peninsula and dissipated on June 23. The storm caused $3 billion in damage and killed 128 people directly, 24 in New York State. The only higher death toll was in Pennsylvania. Agnes dropped 28.1 trillion gallons of water in her nine-day career. The name Agnes was retired in 1973. While the Southern Tier of the state bore the brunt of

Agnes's fury, Onondaga County and the rest of Central New York were not spared from the storm. It was estimated at the time that the storm dumped over a foot of rain in the Syracuse area. President Nixon asked Congress for $100 million in disaster relief. Congress appropriated twice that amount.

No waterway in Onondaga County was spared from flooding. Even the Barge Canal endured the onslaught of Agnes. Residents of Foster Road in Clay were evacuated after flooding that left over a foot of water in basements when the order to evacuate came. The Finger Lakes mitigated the damage that Onondaga County saw. According to Donald Hewitson, regional waterways engineer for the New York State Department of Transportation at the time, the lakes acted as a buffer and prevented the devastation seen in the Southern Tier from occurring in Onondaga County and the surrounding area. In Cicero, New York, Air National Guardsmen evacuated citizens living on the southern shore of Oneida Lake from their homes when flooding swelled Muskrat Bay and covered Beach, Polar Beach and Long Point Roads. Sections of the Erie Canal and the entire Oswego Canal were closed for weeks due to the flooding Agnes brought. Damages in the city of Syracuse came to $20,000, mostly from flooded cellars and sewers backing up.

The southern shore of Oneida Lake still floods in spring after heavy snows. After the blizzard of 1993, several residents suffered damage, and it is an almost annual occurrence that the area of Route 298 known as Rattlesnake Gulch floods. If there is a lesson to be learned from all this watery havoc the county has seen, it is to be prepared and know that no one is immune to flooding. There will be more floods in Onondaga County. It remains to be seen if we have learned the lessons from our previous deluges.

8

LET IT SNOW, LET IT SNOW, LET IT SNOW

Our Notable Blizzards

If you live in Onondaga County you may be assured of two things: you will see road construction from April to November and you will definitely see snow. In fact, our average annual snowfall is approximately ten feet. There is a strong possibility if you live here for any time at all you will also see a blizzard. The cities of Upstate New York vie for the Golden Snowball award. The award goes to the city with the most snowfall for a given season. Usually the prize goes to either Buffalo on the shores of Lake Erie or Syracuse. Occasionally, Rochester, on the shores of Lake Ontario, pulls off an upset and earns the award. Needless to say, it helps to like snow if you're going to live in the area.

We're used to snow in Onondaga County, but sometimes it gets to be overwhelming. We always seem to muddle through. There are blizzards that give the kids a day or two of unexpected vacation from their books. They're inconvenient but nothing to get excited about. Every so often, though, the region gets a snowstorm worthy of not only the name but enduring stories. There was the blizzard of 1888 that prompted future president Theodore Roosevelt to leave a scathing note for a business associate because the man was not at his place of business. The blizzard of 1925 stalled trains and paralyzed the city of Syracuse. Snow buried much of Upstate New York in 1966, giving children and their parents an extended vacation. Snowfall was measured in feet and could have easily been measured in yards. The year 1977 brought another wintery onslaught to Onondaga County. Sixteen years later, three storms would merge to cripple Onondaga County with a

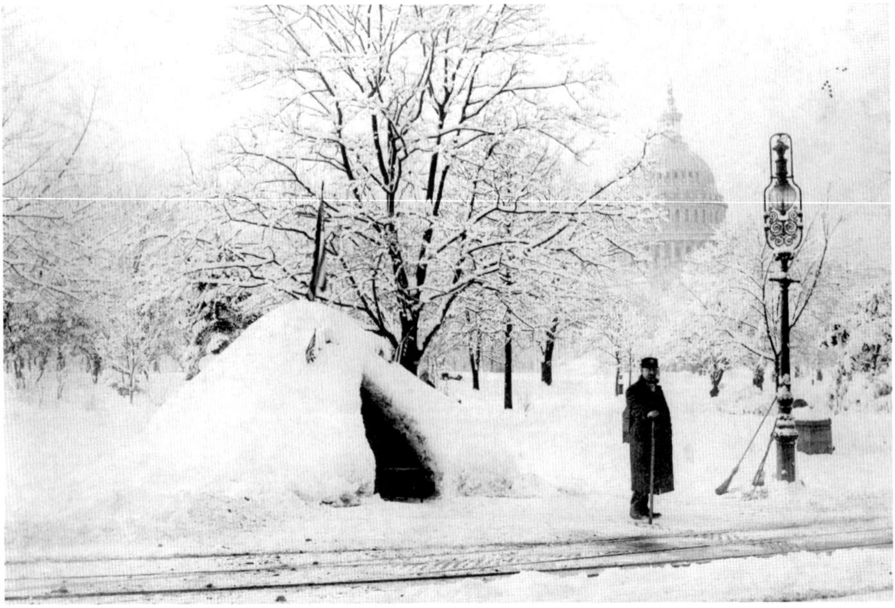

Washington, D.C., during the great blizzard of 1888. The snowstorm paralyzed Onondaga County. It wouldn't be the last time. *Courtesy of the Library of Congress.*

snowstorm punishing enough to be as legendary as the other three. All four storms would have a dramatic and profound effect on the county.

It was called the Great White Hurricane, and it lived up to the name. The great blizzard of 1888 paralyzed the East Coast from the Chesapeake Bay to the Canadian Maritime provinces. The fearsome storm killed four hundred people and caused $25 million in damage, the equivalent of $660 million today. The blizzard began on March 11 and lasted three days. Travel by road or rail was impossible (or at least for everybody but Theodore Roosevelt). The blizzard dumped between four and five feet on Onondaga County. The blizzard of 1888 set the benchmark for the blizzards to follow. The brutal nor'easter buried three-story houses under massive drifts, one of which measured fifty-two feet high.

Onondaga County and Central New York suffered another blizzard on February 12, 1892. Temperatures dropped eighteen degrees. The *Syracuse Evening Herald* stated, "When a storm shatters the windows of a train, you may count on it being a strong one." The newspaper's headline for the event was succinct, stating simply, "It is a very bad storm." No fatalities were reported, but rail and road traffic was disrupted.

By 1925, Onondaga County roads were conveying not only horse and buggy and trolleys, but automobiles as well. Year by year, the price of what was once seen as a toy for the wealthy was becoming more available to the average man. The automobile may be perhaps the biggest boon to mankind—and possibly its biggest curse. Those will be questions for future historians to ponder. It is undeniable that cars added a new dimension to snow removal in Onondaga County. These days, Onondaga County and the City of Syracuse can be justly proud of their capability in snow removal, but it wasn't always so. County citizens have always coped with the snowstorms that on occasion crippled travel in the area. Railroads developed plows to push snow off the tracks. There was no incentive to do so when the automobile was seen as a plaything for the affluent. The blizzard of 1925 changed that.

Like many other places, there isn't much that can't be made into a political football in Onondaga County, including snow.[30] The blizzard of 1925 was no exception. The storm delivered a crippling blow to the area on February 2, as the city was digging out from a hefty snowfall that preceded it by three days. Syracuse was forced to ask for county assistance in removing the snow. Traffic throughout the county was paralyzed. It was reported that burglaries decreased during the storm. The blizzard of 1925 taught Syracuse and Onondaga County lessons in dealing with snow removal. In the ensuing years, the lessons learned would be refined, and new problems would arise.

It was a schoolboy's dream! Really, who wouldn't relish a week off for free from school? The official name for the storm is the North America Blizzard of 1966, and its effects were felt throughout the United States and Canada east of the Rocky Mountains. For people of my generation, it is the storm of the century even if the blizzard of 1993 was more damaging. Ironically, the seasonal snowfall for the 1965–66 season was about average at 118.8 inches.

The storm began as a classic nor'easter and cycled over the Great Lakes generating lake-effect snows that would drop over 3 feet of snow on Syracuse. On January 28, the temperatures were brutally cold, with Syracuse recording a high temperature of zero degrees Fahrenheit, thirty-one degrees lower that the normal temperature for that time of year. The storm dumped 44.3 inches of snow, over 3.5 feet on Syracuse in two days. Over 55 inches of snow fell on the area in three days.[31] Roads throughout Onondaga County and New York State were impassable. The New York State Thruway was closed from Albany to Buffalo, and Hancock International Airport was closed with no flights in or out over the course of the storm.

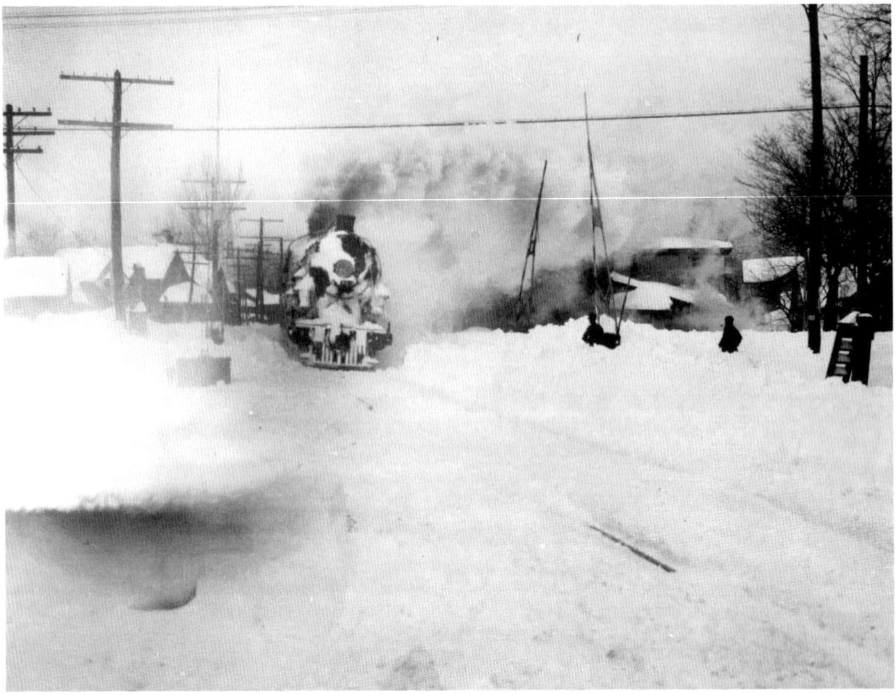

The 1120 Flyer steams through Liverpool during the blizzard of 1925. *Courtesy of the Liverpool Public Library, from the Schuelke Collection.*

Almost immediately, herculean efforts to clean the streets and get things back to normal began. At least seven people died during the storm, and an eighteen-year-old Baldwinsville youth disappeared while trying to visit his girlfriend. His body was found on February 13 by a search party consisting of his father and family friends a scant five hundred yards from safety. Four of the dead were children killed in a Nedrow house fire.

Millions of dollars in business losses for the area resulted from the storm. Carrier and General Electric, both multinational corporations and two of the biggest employers in the region, shut their doors for several days during the storm. County businesses lost an estimated $5 million a day due to the weather-related shutdown. The City of Syracuse alone spent over $217,000 clearing the accumulations of snow. In rural areas of Central New York, some of the stranded were rescued by helicopter. Snowmobile owners were pressed into service to get the ill or injured to hospitals, and food and medical supplies were dropped by parachute to some in need. The blizzard of 1966 proved again that weather can wreak havoc on the area in short order. Weather forecasting did not have the tools to predict storms days in advance as they do now.

Two absolute truths about Onondaga County in the wintertime are that we will get snow, and invariably, the blizzard of 1993 will get mentioned if the threat of another blizzard pops up. As I stated earlier, there are still debates as to which storm was worse. If we just look at statistics, the "Storm of the Century," as the 1993 blizzard was called, takes the dubious honor of worst storm. For impact, the 1966 blizzard wins. Ironically, though the 1993 blizzard dumped more snow on the area in a quicker amount of time, better technology allowed the State of New York to declare a state of emergency for the state and Onondaga County to marshal resources to combat the snow. The March storm was the first time the emergency system was activated statewide. Digging out from the 1993 blizzard cost county taxpayers $350,000 as of March 18, 1993, and the bills weren't all calculated. Most of the expense was wrapped up in overtime for the crews clearing the snow. In a pleasantly unusual turn, no injuries were reported in the Central New York area. The storm shattered records for snowfall in a twenty-four-hour period and the record for snowfall in the month of March.

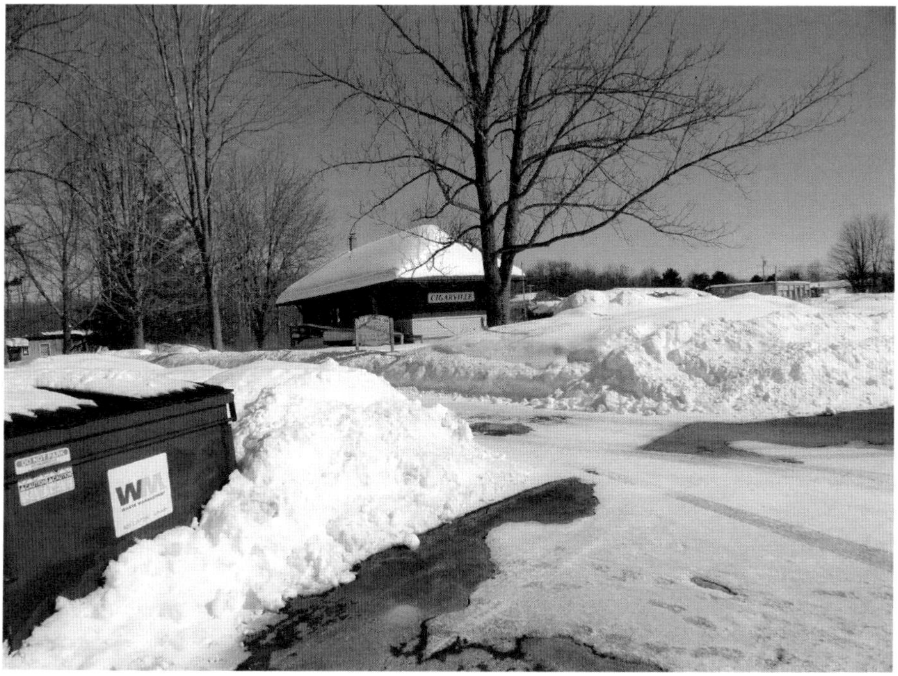

Cigarville Station in Clay. This is not an unusual amount of snow for Onondaga County. *Neil K. MacMillan.*

There was damage. A section of the Penn-Can mall roof collapsed under the weight of the snow from the blizzard. With the thaw, basements did flood, and it took time to unbury cars. Unlike the blizzard of 1966, the weekend storm did not seriously affect the manufacturing sector in Syracuse. For some businesses, like the area ski resorts, the blizzard proved to be beneficial. The additional snow added to their coffers in the form of more hours on the slopes for avid skiers.

As we move further into the twenty-first century, there will be more snow in Onondaga County's future. With each storm, we get better at predicting the storms and handling the snow that results. The lesson remains that for all our technology, the weather can still throw us a wicked curve ball.

9
WHAT A BLAST!
The Explosions that Rocked Our World

Some catastrophes build with time. Explosions don't. There is almost always an underlying cause that may take time to develop, but the result is brutally quick and devastating. Most are caused by a malfunction or error—some are caused for more sinister reasons. We can legislate against explosions—or try—yet they still occur. The problem is, there will always be under-maintained equipment. Human beings will prove time after time that we are not perfect, and sometimes, some of us prove that we can be evil. I do not include terrorist attacks in this chapter. Those are acts of war, and I covered the Pan-Am attack in another chapter.

Night settled over Syracuse on August 20, 1841. As businesses verged on closing and people were gearing down to retire for the evening, church bells began to ring. Church bells chiming on a Friday evening meant fire. The conflagration began as an apparently modest fire in Charles Goings's carpentry shop on Church Street, and Syracuse's volunteer fire companies responded with enthusiasm. Firemen dragging their hand pumper were urged forward by their foreman's cries of "Let her go, Boys!" and "Let her jump, Boys!" A crowd gathered to watch the firemen battle the blaze. As the intrepid firefighters went about their duties, the mass compressed in on the scene. Someone in the press of people called out, "There's powder in the building!" Up to this point, the fire seemed to be under control. Some of the crowd moved back, but most held their ground—safe in the knowledge that it was a violation of village ordinances to store gunpowder in the village. It was a bad career choice. The building disappeared with

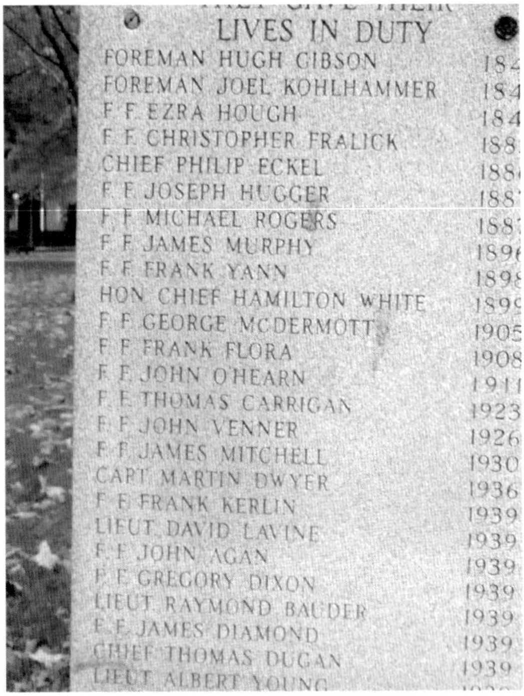

Right: Roll call of the fallen. The first names are those of firefighters killed during the gunpowder explosion. *Neil K. MacMillan.*

Below: Coarse salt mills in Salt Spring burn as firefighters try to extinguish the blaze. *Courtesy of the Liverpool Public Library, from the Schuelke Collection.*

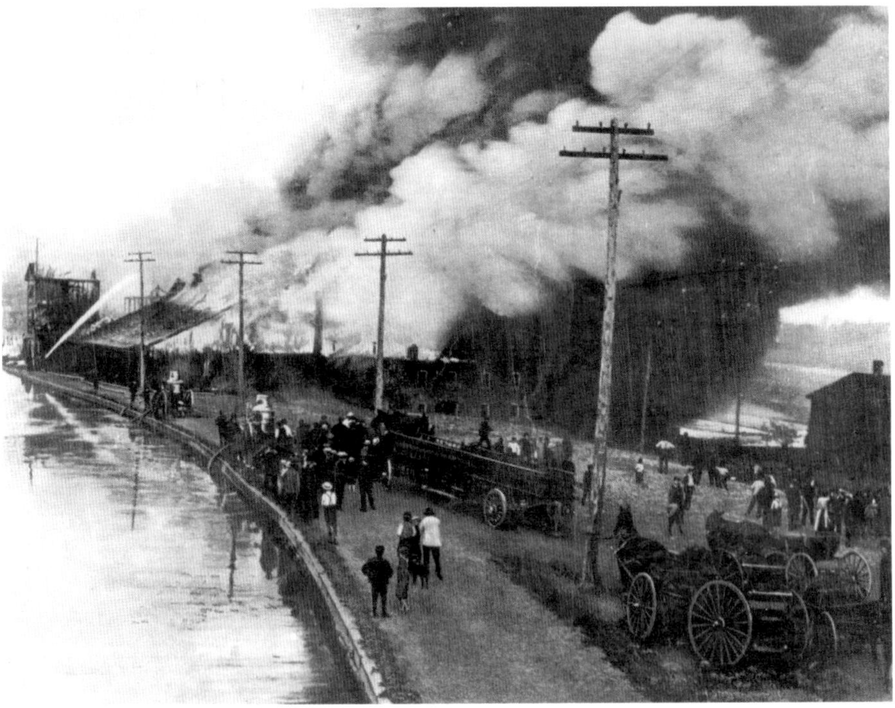

a resounding thunderclap and a brilliant flash. Flying timbers coursed through the sky, and flames lit the area for five seconds. The blast was felt as far away as Fulton, New York, approximately thirty miles away. At first, some in the outlying towns thought there had been an earthquake.[32] Twenty-six people were killed. More than sixty were horrifically injured. Among the dead were six firemen, including a merchant and foreman of the firefighters, Hugh T. Gibson, who gallantly urged his men forward into the fray to stave off the disaster.

No cause for the blaze was discovered. Bear in mind these were the days before forensic science was dreamed of. However, the newspapers of the day speculated that the owner of the carpentry and joiner shop, Charles Goings, had made enemies in his business dealings. There might have been truth to the newspapers' assertions. The gentleman who delivered the gunpowder to Goings's shop reputedly told one of the neighbors about the powder. The deliveryman, Ingall, also cautioned the neighbor not to say anything about the powder after recommending the neighbor move. It is likely that in a similar situation today Mr. Ingall would have been answering some pointed questions from the authorities. Was the inferno arson? That is a possibility and one that in 1841 would have been much easier to accomplish without fear of detection than it would be today. If so, who stands out? After 175 years, the prime suspect would be Ingall, but only because of his cryptic warning. Did one of Goings's enemies take advantage of the easy kindling in the form of carpentry shavings? At this juncture, we will never know.

Charles Goings, a Mr. Malcom and a Mr. Hudson, the owner of the gunpowder, were found guilty of negligence and violating a village ordinance. The disaster was the worst in the county's history until the explosion at Split Rock seventy-seven years later.[33] There is a plaque commemorating the disaster at the site.

If there is any disaster that still scars Onondaga County, it is the explosion at Split Rock Quarry on July 2, 1918. The explosion still reverberates in our collective psyches, and the old quarry is rumored to be haunted because of the explosion.[34] In 1914, Archduke Franz Ferdinand and his wife, Sophie, were murdered by Bosnian nationalist Gavrilo Princip. The assassinations propelled Europe into a war that only needed a spark to set off conflict. Years of vying for colonies and military power turned Europe into a hotbed of political and diplomatic intrigue.

Split Rock Quarry started life providing limestone for the Semet-Solvay Corporation. Limestone was the basis of the firms manufacturing process

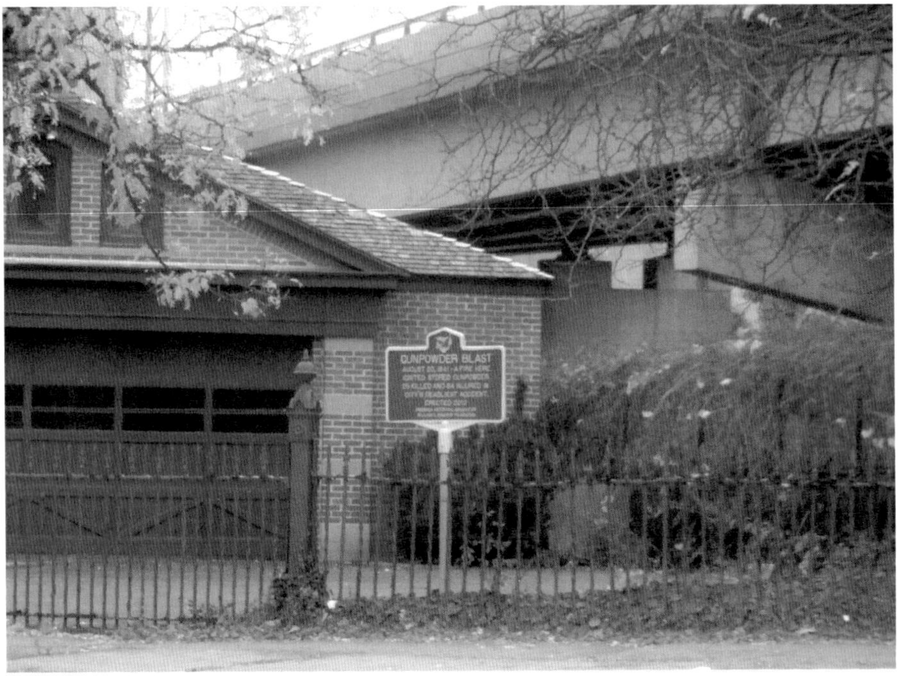

The site of the 1841 gunpowder explosion. Twenty-six people were killed and over sixty maimed when Charles Goings's joinery burned. *Neil K. MacMillan.*

for potash, but by 1907, the quality limestone needed for the process had been played out. The quarry stood idle for the next seven years. When Europe went to war in July 1914, the German Empire smashed through Belgium to deliver a quick death blow to France. France held, and the United Kingdom declared war on Germany. Almost as soon as the guns began firing, both sides began seeking guns, ammunition and military supplies from the United States. While the United States was neutral, the Semet-Solvay Corporation was decidedly for the Triple Entente.[35] The company's founders, the Solvay brothers, were Belgian immigrants to the United States and quickly began taking orders from the countries aligned against Germany. The company began manufacturing high explosives, notably TNT. Split Rock Quarry was quickly converted to a munitions factory.

By the evening of July 2, 1918, safety procedures had been well established for the munitions factory. In the event of a fire, employees were to evacuate. At 8:40 p.m., a fire started in a mixer in Plant One. The mixer was used for combining chemicals in the manufacture of TNT. Though the employees were supposed to evacuate, all of them knew America was at war and needed the factory's output. They also knew that if the factory were destroyed, they would lose their well-paying jobs. The men manned fire hoses and began combatting the fire. Sometimes heroism is rewarded. This wasn't one of those times. Around 9:30 p.m., the employees were making headway against the fire when the water pressure dropped. An explosion shook the quarry.

The blast shook buildings for miles around and partially damaged the school in the hamlet of Split Rock. Seventy people died in the blast.[36] To date, it is the most devastating explosion in the county's history. Almost as soon as the recovery efforts began, the investigation started, and rumors began to circulate. Onondaga County District Attorney Walrath concluded after several days of testimony and reviewing the evidence that a gear in the number one mixer in Plant One had overheated, causing the fire and subsequent explosion. Walrath also determined that the water pressure drop contributed to the blast and the water pipes and pumping equipment had not been adequately maintained. But was the blast merely a tragic accident?

In the midst of World War I, rumors abounded that the explosion was a result of sabotage by German agents or sympathizers. Before discounting the rumors and upholding District Attorney Walrath's conclusions, there

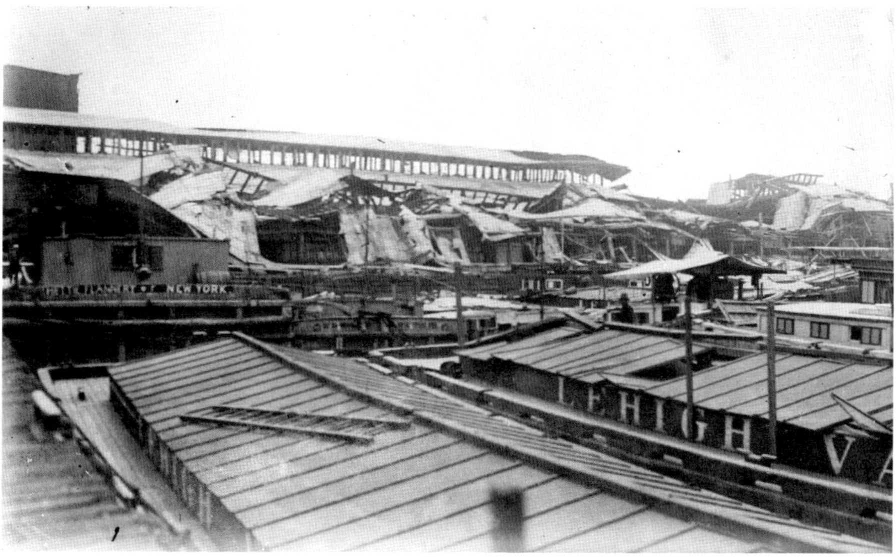

The damage wrought by the Split Rock explosion was never repaired. Four months after the explosion World War I ended, the quarry was sold to New York State. *Courtesy of the Library of Congress.*

are several facts to consider. Why was detonator cord found near the site of the explosion when Split Rock Quarry ceased quarrying eleven years earlier and the site was used as a munitions factory and storage depot? Then there are the expended rifle cartridges found at the site years later. The guard force at Split Rock carried rifles chambered for what was even then a rare cartridge: the Remington .32-70. It is known that the Semet-Solvay guards carried that caliber of weapon both at the main factory in Solvay and at Split Rock Quarry. It is also fact that the German government threatened to destroy the facility at Split Rock. The Imperial German government had already sabotaged the Black Tom munitions plant in New Jersey.

The flip side of this particular coin is also compelling. After the end of the Great War, Semet-Solvay sold the quarry to the State of New York, which used it as a Department of Transportation facility until closing it in 1980. There was talk of turning it into a park with hiking trails and historical markers. So far, nothing has come of the plan. The old quarry has been used as an impromptu shooting range and teenage party spot. Did the shells come from a local marksman plinking at tin cans? It is possible, though the cartridge has not been manufactured since 1907 and if any weapons chambered for it are still extant in firing condition then their owners are making their bullets. Did Walrath get it wrong? Perhaps. Semet-Solvay attempted to keep the investigation findings within the company. With the deaths of over fifty people, that was never a serious consideration. It is my opinion that this was to suppress the fact that there had been neglect in maintaining the water system for the quarry and out of concern that the company would suffer from the resulting publicity.

It has been ninety-nine years since the explosion, and as historians uncover more about the blast, more questions seem to pop up. Accident or sabotage, the Split Rock explosion remains the county's worst disaster to date. We can only hope nothing happens to top the current record.

Almost from the earliest days of Onondaga County's history, Onondaga Lake has been one of the area's favored recreation spots. On any given day during the summer months and well into autumn, boats ply the waters of the lake. It was no different in 1885. In those days, several steam vessels gave water tours of the area. Among these was the steamer *John Greenway Jr.* Captain Charles Kinne had commanded the vessel for several years and was reputed as a competent but stubborn captain. Yet the vessel had problems. The ship's engineer, Kinne's nephew Antonio, was inexperienced, and it was rumored that the ship's boiler was in poor operating condition. The boiler had been tested at 125 pounds per square inch (psi) pressure. On May

23, 1885, the *Greenway* was preparing for another day on the lake. Boiler pressure was at 95 psi when the vessel left Geddes with twenty passengers. An explosion ripped through the vessel, blowing out the boiler flue and chimney. Jacob Grassman suffered burns to his hands. Captain Kinne's entire body was scalded by the escaping steam, and Antonio was severely burned. Captain Kinne succumbed to his injuries the next morning, and Antonio perished later. The vessel had been noted to be unsafe, something Captain Kinne stoutly denied. Perhaps he strove to protect his nephew's reputation. Whatever the reason for his denial of the *Greenway*'s problems, his hubris proved fatal.

Fifteen years later, another vessel would have boiler problems with fatal consequence. On July 24, 1900, the yacht *Trilby* set off on the Oswego Canal for a pleasure cruise. The yacht was under the command of Fred Spink, the owner and captain. Aboard were his family and Mr. and Mrs. Jay Kelsey, family friends. As the *Trilby* approached Syracuse, a boiler flue loosened. The resulting explosion blew open the fire box door, scattering hot coals on the passengers as escaping steam scalded them. As horrified spectators along the Oswego Canal watched, seven people were severely

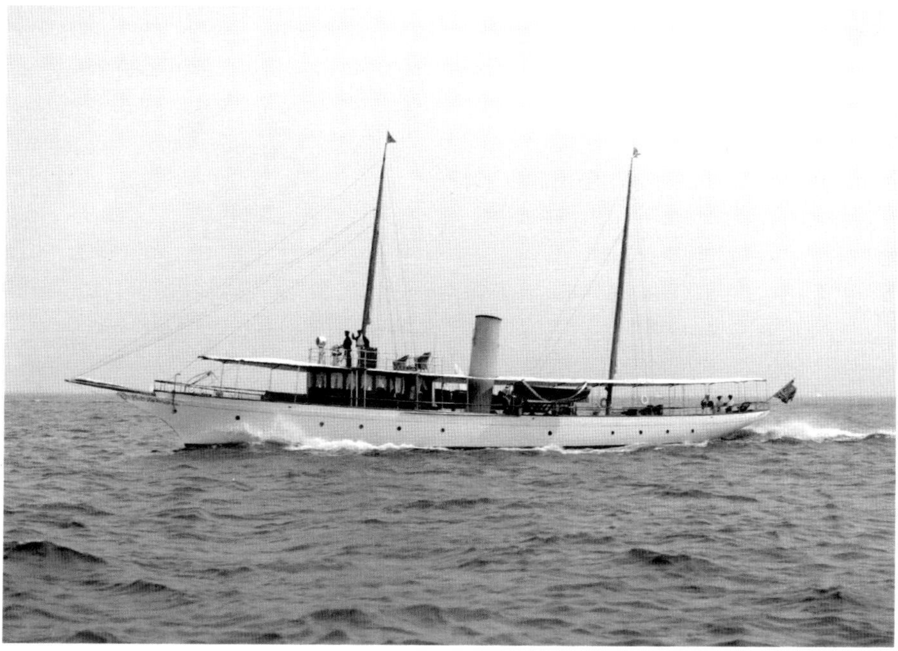

The steam yacht *Akela* was similar to the *Trilby*, which suffered a boiler explosion in 1900. *Courtesy of the Library of Congress.*

injured. In an instant, the explosion ripped through the Spink family. All four of the Spink children died of their injuries. Iva and Gladys died at 7:40 p.m. that evening.[37] Sister Fern died a few hours later, and brother Herald passed away in the early hours of July 25. Mrs. Spink was severely burned, and the captain suffered burns to his arm and hand. Mrs. Kelsey suffered burns to her lower legs. Investigators quickly concluded that the boiler's faulty construction led to the blast. The ship builder had died, so there were no charges filed in the incident. Construction shortcuts turned a Sunday cruise into a horrific tragedy.

What are the lessons of the explosions mentioned here? The biggest lesson is that as a species, we are not infallible. We can propose and pass legislation to prevent catastrophe, but there will always be those who feel a need to circumvent those rules. The Goings fire and explosion would be unknown today if Mr. Goings had not stored twenty-six barrels of gunpowder in his shop in violation of village ordinances. People will take shortcuts when it comes to their jobs. Had Captain Kinne engaged a competent ship's engineer instead of giving his nephew a job, both men might have lived well into old age. We can hope that the days of devastating explosions are behind Onondaga County, but let us also keep a vigilant eye.

10

WRECKAGE ON THE PARKWAY

The Onondaga Lake Parkway Overpass and Large Vehicles

Onondaga Lake Parkway is a scenic drive along the eastern shore of Onondaga Lake. The four-lane roadway gives quick access to the village of Liverpool from the outskirts of Syracuse and passes sites that were important in the county's early history, including Sainte Marie among the Iroquois and the Salt Museum, honoring the industry that gave Syracuse its nickname: the "Salt City." The parkway relieved traffic on Old Liverpool Road, but the railroad overpass proved to be a problem for some drivers. The overpass was built in the day when Onondaga Lake Parkway was part of the Oswego Canal. Low-slung canalboats had no problem with the overpass. The railroad overpass is ten feet, nine inches from the road surface. It poses no problem for cars, but trucks are another story altogether.

Just as we know we'll get snow and lots of it, we know that someone will try to fit a thirteen-foot-tall truck trailer under an underpass that is less than eleven feet tall. Since the parkway was opened in the 1930s, truck drivers have misjudged the height of the overpass. Most of the time, traffic is gridlocked for hours and the driver might suffer minor injuries—but the overpass has claimed lives. Even before Onondaga Lake Parkway opened, it was the center of controversy. In 1929, the New York State Department of Public Works opposed the original plan, as drivers would not be able to see the lake. The state plan wanted the proposed boulevard to run along the lake. Since the county could not complete the proposed work without state money, the plan to run the route through Galeville was abandoned and merged with the plan for the Onondaga Lake Parkway.

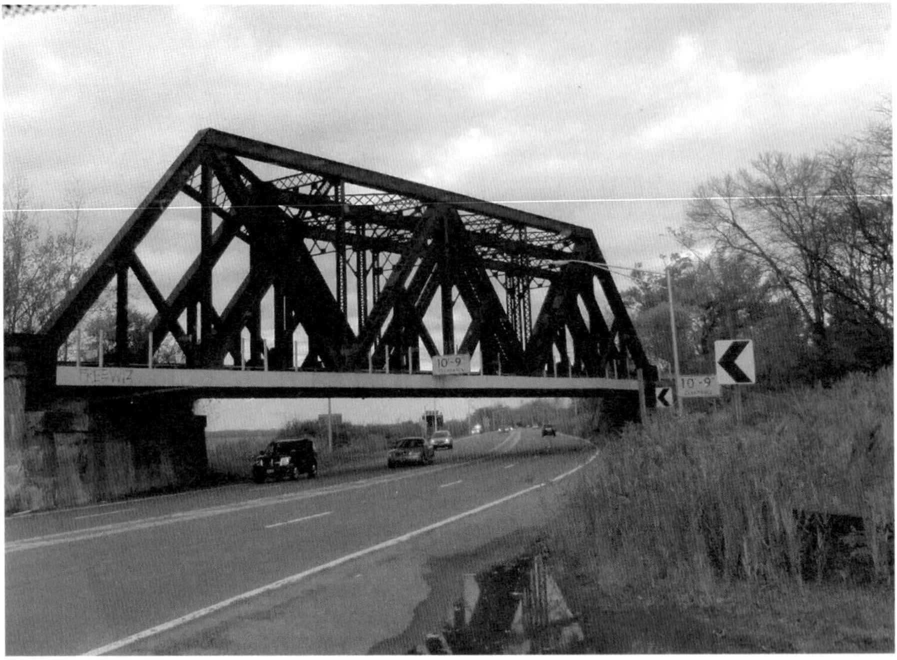

The notorious railway bridge spanning Onondaga Lake Parkway. In spite of myriad traffic signs and devices warning of the low bridge, which is only ten feet, nine inches high, almost every year someone runs into it. In 2010, a Megabus double-decker bus slammed into the bridge, killing four and injuring ten. *Neil K. MacMillan.*

The controversy continues to this day. Every time a truck slams into the overpass, the newspaper is flooded with letters touting the cure for the parkway's problem. Onondaga Lake Parkway is a beautiful drive through one of the county's richest historic areas. It also has a darker history.

On March 1, 1958, Rhode Island trucker James Viera slammed into the overpass. When the trailer hit the bridge, the truck flipped on its side. Viera suffered bruising on his back, hips and shoulder; he was not the first to misjudge the overpass and wouldn't be the last. By 1963, citizens and truckers were clamoring for something to be done about the overpass. Four trucks slammed into the railroad overpass that year. Among the proposals were ideas to raise the overpass, lower the road, divert all truck traffic to Old Liverpool Road and put up more warning signs and traffic devices. County executive John Mulroy was in favor of more and better warning devices. He saw trying to raise the overpass or lowering the road as impractical. State assemblyman John H. Terry proposed legislation to require truck cabs to have the height of their trailers prominently displayed in the cab. Among the hazards listed of the parkway were poor lighting and the overpass. One intrepid driver was sought by state police on September 12, 1967, after he slammed into the overpass and kept going. The crash ripped the top of his trailer off, tying up traffic on Onondaga Lake Parkway.

A truck driver was charged with exceeding bridge clearance and driving while intoxicated after his rig struck the overpass on Friday, August 16, 1968. The driver suffered cuts to his hands. He was treated at Memorial Hospital and then arrested. Onondaga Lake Parkway was closed for four hours while the wreckage was being removed. The task was made more difficult by the fuel the trucker was carrying. The Middleton driver was held in lieu of a $100 bail. Given his cargo, the crash could have been much worse.

By 1972, sixteen signs warned drivers of the low bridge. Between 1962 and 1972, the number of signs had increased by fourteen, yet trucks kept ramming the overpass. Theories were offered at the time that driver fatigue played a part in the mishaps or that some of the drivers were functionally illiterate, a position that prompted heated denials in the editorial columns of the day. Before the increased number of signs on the parkway, three to four trucks a year smacked into the overpass. The New York State Department of Transportation believed the increase in signs mitigated the problem since there had been no trucks crashing into the overpass for twelve months after the installation. August 1972 proved them wrong when two trucks struck the overpass. One driver who struck the bridge asked after her truck and cargo and seemed little fazed that she would not only face a $100 fine but also accrue the costs of the cleanup and any repair to the bridge.

A load of asparagus littered the parkway after a Weehawken, New Jersey truck driver failed to heed the plethora of signs warning about the low overpass on April 8, 1977. The trucking-eating overpass sheared the top of the cab and trailer. The driver escaped with minor injuries. The parkway isn't the only road with an overpass problem. The rail overpass on North Midler Avenue has also been struck, but the parkway is notorious, while Midler Avenue sees very few accidents with its overpass.

The 1980s saw little mitigation of the problem. A Baldwinsville truck driver struck the Onondaga Lake Parkway overpass while transporting a load of frozen food. The crash separated the trailer from the tractor. Department of Health officials held the frozen food to determine if any of it could be safely salvaged. The driver was cited for failure to obey a traffic device. He should have known better; unlike many who struck the railroad trestle, he was from the area and had heard the stories about the bridge. In 1982, the

Village of Liverpool bought a fire truck designed to fit under the overpass. It allowed firefighters to respond with better speed to calls in the area. A driver delivering furniture struck the bridge and turned over his truck in 1986.

An Oklahoma City trucker found that his rig would fit under the bridge if he didn't mind the top being sheared off. The truck spilled a load of inner tubes on the parkway, and while the truck made it through the overpass, the damage resulted in the parkway being closed for several hours and the driver charged with failure to obey a traffic device. The year 1987 saw the overpass reach new heights of notoriety. Eight trucks bashed into the overpass that year. Among them was a fuel truck that spilled gallons of diesel fuel over the parkway, requiring a chemical cleanup team to remove the oil spill. A Virginia driver stated he didn't see the signs. Another driver didn't have the excuse of not being a local when his truck struck the bridge and spilled a forklift onto the parkway. A state DOT employee maintained that the drivers were the problem: "It's the drivers themselves. The lights and signs are adequate. Truckers are going too fast and when they realize it, it's too late." Perhaps the employee was right. Speed is essential in the trucking industry to ensure the cargo gets where it needs to be on time. One driver called the bridge a "booby trap." At the time the comment was made, the Onondaga Lake Parkway had been serving the area for over fifty years. The insidious booby trap had been in service even longer, originally spanning the Oswego Canal. The bridge was painted fluorescent orange in an effort to make it more visible to truckers in early 1988.

Assemblyman Michael Bragman inquired into the cost of either lowering the road, raising the overpass or rerouting rail traffic so the trestle could be eliminated entirely. Reflective diamonds were installed on the overpass in July 1988. New York State quashed plans to reroute truck traffic onto Old Liverpool Road or Route 57 after residents of the area complained that the increased traffic would be detrimental to their neighborhood. State DOT officials claimed that the traffic increase would be minimal. A driver delivering liquor to a local distributor struck the overpass on March 17, 1989. It took several hours to clean up the accident. One man suggested hanging bells to alert truck drivers of the low bridge. The man suggesting the bells stated it was only a matter of time before someone was killed. A Florida driver was treated for back pain after his rig slammed into the overpass. The crash was the fourth of 1989 involving the overpass. The truck spilled frozen orange juice on the parkway.

A Texas truck driver, who was a trainee for his company at the time, crashed into the bridge in January 1992. His trainer was also in the cab during the incident. The driver stated to police, "Yeah, I saw the signs." They were hauling corrosive powders, but fortunately, the material did not spill onto the road. Passing motorists honked their horns and offered unsolicited commentary on the driver's vehicle-handling skills. The driver was ticketed for failure to heed a traffic device. Another driver found out that his beer truck wasn't as small as he thought it was. The Virginian told police he misread the signs. There was a cartoon extant in the early 1990s here showing a policeman ticketing a truck driver. The cop asks the driver if he is stuck. The driver replies in the cartoon, "Nah, I'm delivering a bridge." As the technology and signage meant to decrease or eliminate trucks from ramming the trestle increased, it seemed like the rate of trucks getting stuck kept pace. From 1986 to 1995, twenty-two truck crashed into the overpass, causing county legislator Bill Sanford to call legislation to curb the problem. Sanford stated in his speech that the greatest sport on Onondaga Lake Parkway was watching trucks slam into the bridge. Five of those accidents occurred in 1995. The overpass was sated from July 1995 to 1997.

It was only a matter of time before the bridge claimed a life, and when it did, it did so with a vengeance. Four people were killed and twenty people injured when a Megabus double-decker bus slammed into the bridge on September 11, 2010. The driver had taken a wrong turn and was trying to reach the Regional Transportation Center. The accident happened on the ninth anniversary of the terrorist attacks on New York, Washington and Shanksville, Pennsylvania. Ironically, it was not the driver's first time driving in Syracuse. He had driven the same route nine times previously without incident. The driver stated he was lost and unsure of where he was. Emergency responders were quickly on the scene, and it is a tribute to their skill that only four people died. The death toll could have been much higher. Passengers helped others get out of the stricken coach. Lawsuits ran into the millions of dollars, although all parties agreed to not disclose the specific amount awarded in those cases that were settled. It was claimed that the driver was using his personal GPS rather than the GPS unit installed on the bus. The personal GPS did not take into account the height of bridges in calculating the route for the driver. In the wake of the tragedy, area residents again put forth ideas on how to mitigate the accidents at the trestle.

Several ideas have repeatedly been offered as a solution to Onondaga Lake Parkway's problem. One of the most frequently mentioned is to lower the road. This is one of the ideas that was seen as impractical. Onondaga Lake Parkway currently is approximately at the same level as the lake. Lowering the road in the area of the overpass would not only be costly but could also result in flooding and excessive freezing during cold and wet seasons. Money also plays a factor in raising the trestle. The grade would have to be gradually increased on both sides of the trestle to enable trains to still use the tracks. The cost would be even more than lowering the road. A third alternative is to do away with the overpass. As with the second solution, the railroad, Conrail, would have to agree to the plan. It is unlikely to agree to either plan. Removing the overpass would force rail traffic to take a circuitous route via Albany to get to points in the North Country. Signs, bells, electronic devices and lights have all been tried, with less than stellar results. The only surefire way to stop truck and bus traffic from slamming into the overpass is to ban all bus and truck traffic from the parkway and reroute Old Liverpool Road as Route 370. So far, area residents have vigorously opposed the idea, claiming it would result in an excessive increase in traffic.

Of the ideas mentioned above, I believe rerouting all truck and bus traffic to either Old Liverpool Road or Route 57 is the right one. Until the area's residents get behind the idea or the state makes it happen in spite of objections, trucks and buses will continue to smack the overpass. Hopefully, no one else will die or be injured when it happens.

11
THE DAY THE BOTTOM FELL OUT

The Erie Canal Drains

The Erie Canal was once the lifeblood of not only Onondaga County but also the entirety of Upstate New York. It can be rightly said that the canal opened an era of economic boom for Upstate New York and was instrumental in opening the West to settlement. From the day the Erie Canal opened, its fate was intertwined with that of Onondaga County. In the days before the canal, travel was by roads cut through dense forest or by rivers in canoes portaged in places where the water was too shallow for them to float. The Erie Canal was not an original idea; indeed, canals had been used in Pharaonic Egypt. "Clinton's Folly," as the Erie Canal was derisively known by Dewitt Clinton's detractors, would link the city of New York and the Atlantic Seaboard with the Great Lakes. Clinton's vision was energetic and far-reaching. He gambled that the technology of the early nineteenth century could overcome the myriad obstacles the topography of New York State offered. Governor Clinton was right.

When the canal was completed in 1825, Onondaga County goods could move east to New York City and west to the Mississippi River. A journey that took weeks was shortened to days. The Erie Canal opened trade and allowed New York State's economy to grow into a powerhouse. Tolls collected on the canal paid for its construction in the first year. In addition, the canal brought new ideas to the state and points west. The ethnic makeup of communities all along the canal changed with the influx of immigrants who found jobs building the Erie Canal or working on the various improvement projects during the canal's history.

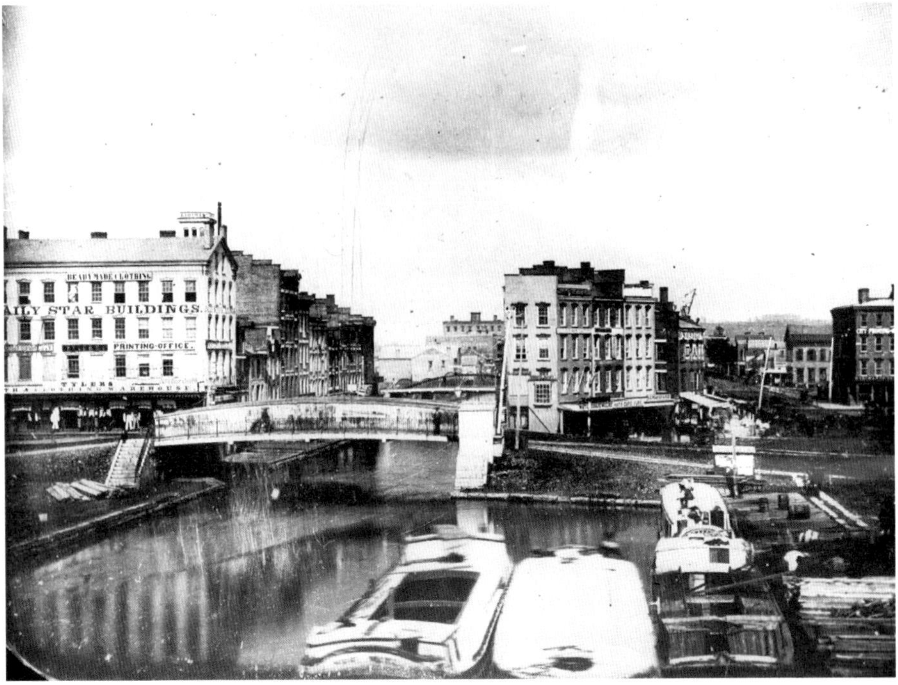

The Erie Canal flowed through downtown Syracuse before the collapse. *Courtesy of the Library of Congress.*

Almost as soon as the canal was completed, it faced competition from an unseen source at the time of its completion. In 1839, Syracuse's first railroad station opened at Vanderbilt Square. As railroads crisscrossed the nation, they became able to ship goods at a price the canal couldn't compete with. By 1900, most goods were being shipped by rail. Trains could get more cargo across the state in hours instead of days. The days of the canal's economic supremacy lasted about a generation. The Erie Canal soldiered on in spite of fierce competition from the railroads. The canal endured because some people were nostalgic, but for the most part, it continued because many people saw the railroad executives as greedy businessmen out to gouge them. There may well have been truth in those assertions before legislative action curbed the power of big businesses to monopolize areas of trade.

Like any aspect of the infrastructure, the Erie Canal needed improvements. One of the first came within two decades of the canal's opening. Traffic was greater than the engineers who designed the canal had anticipated. Between the 1830s and 1862, the Erie Canal was widened and deepened to allow for bigger canalboats and more traffic. It was also discovered that the increased traffic was eroding the clay bottom of the canal. Leaks had to be repaired. In 1838, a culvert was constructed, allowing the Erie Canal to pass over Onondaga Creek.[38] The canal underwent several repairs and improvements in the period between its opening in 1820 to its demise in 1918. Today, parts of the original canal are parks dedicated to the history of the canal and its effect on Upstate New York and Onondaga County.

On July 30, 1907, a keystone in the Onondaga Creek culvert worked its way loose. The collapse of the culvert opened a huge hole in the Erie Canal. The resulting drainage of miles of the canal resulted in the destruction of five canalboats and extensive damage to several buildings along the canal. Amazingly, no one was killed in the calamity. Among the damaged buildings were the Amos block and the Greenway brewery. Both buildings' back walls crumpled into the canal; among the debris were several barrels of flour from the Amos Mill. The Greenway brewery was so badly damaged that it moved operations to the corner of Hiawatha Boulevard and Park Street. The sudden influx of water into Onondaga Creek caused the creek to flow upstream toward the valley section of the city and the hamlet of Nedrow. Repairs took months, and while the canal continued to function after the repairs, the damage to local business was done. Area brewers depended on the canal to ship their product cheaply. When the bottom fell out, they had to pay the railroad's prices. It was a harbinger of things to come. The damaged culvert was repaired by September 1907, but damaged canalboats were unable to be lifted from the canal. They were burned in place, and some timbers are still visible according to local author and news columnist Sean Kirst.[39]

To some, the canal was seen as a necessary evil. It was smelly from years of pollution, and in every city the canal passed through, vice-ridden areas were found. In Syracuse, prostitution and other crimes related to that industry flourished around the area west of Clinton Square. Soiled doves knew where the "canaullers" congregated. Perhaps operating near the city jail and the courthouse wasn't the brightest idea. As Onondaga County surged into the twentieth century, the county's citizens wanted an end to the vice in the areas by the canal. The collapse of the Onondaga Creek culvert helped them in their endeavors. Rail traffic and the new and growing phenomenon of the automobile spelled the canal's economic demise. Why pay canal tolls when you could ship for a flat rate by rail or load your cargo in a truck and ship it on the ever improving roads throughout the state? The state constructed the barge canal, laying it out along natural waterways. Towpaths were no longer

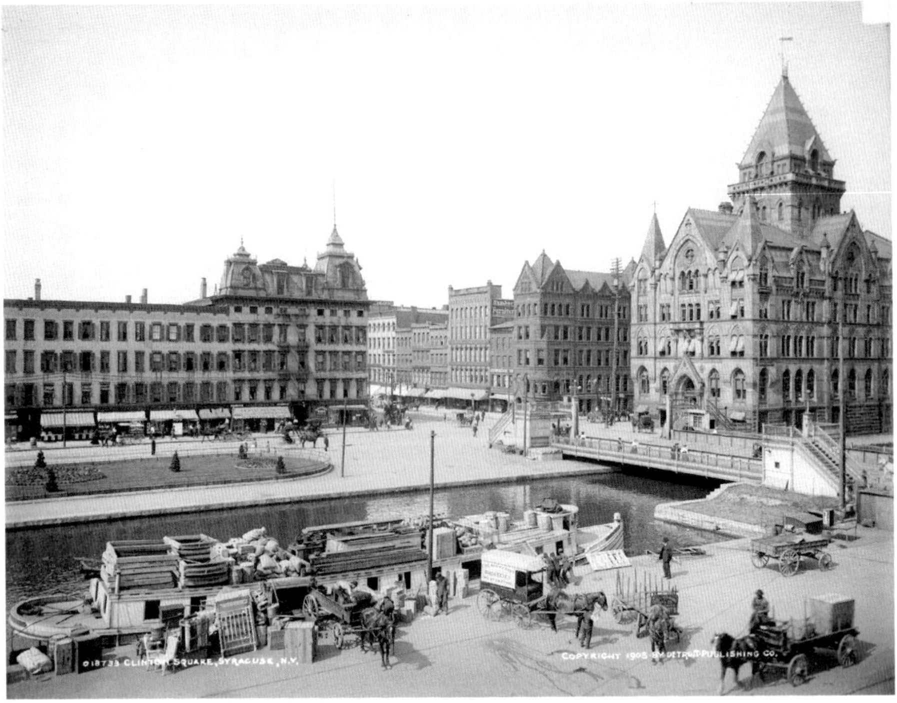

When the Erie Canal drained in 1907, it wrecked vessels and buildings in moments. Though it was repaired, the canal's days were numbered. *Courtesy of the Library of Congress.*

required and nor were "hogees," the boys and young men who led the recalcitrant mules pulling the boats along the towpaths. The last canalboat plied the Erie Canal in Syracuse in 1922. With its passing—perhaps hastened by the collapse of the Onondaga Creek culvert—an era was gone.

The canal was filled in, and one of Onondaga County's major roadways replaced it. In 1923, Erie Boulevard opened, tracing the route of the canal. The sounds of car horns replaced the braying of ill-tempered mules. Speeding cars and trucks are now seen instead of slowly moving canalboats.

The Onondaga Creek culvert is still there. Erie Boulevard passes over it now. The culvert underwent a major renovation in 2012 to repair damage brought about by years of heavy traffic on Erie Boulevard. As of this writing, the culvert is inaccessible to the public. The property on which it sits belongs to a power company, National Grid. Even the much-touted creek walk does not allow a view of the culvert.[40] There are those who question the collapse of the canal being a disaster. Certainly no one died in the collapse, and injuries were limited thanks to quick thinking and

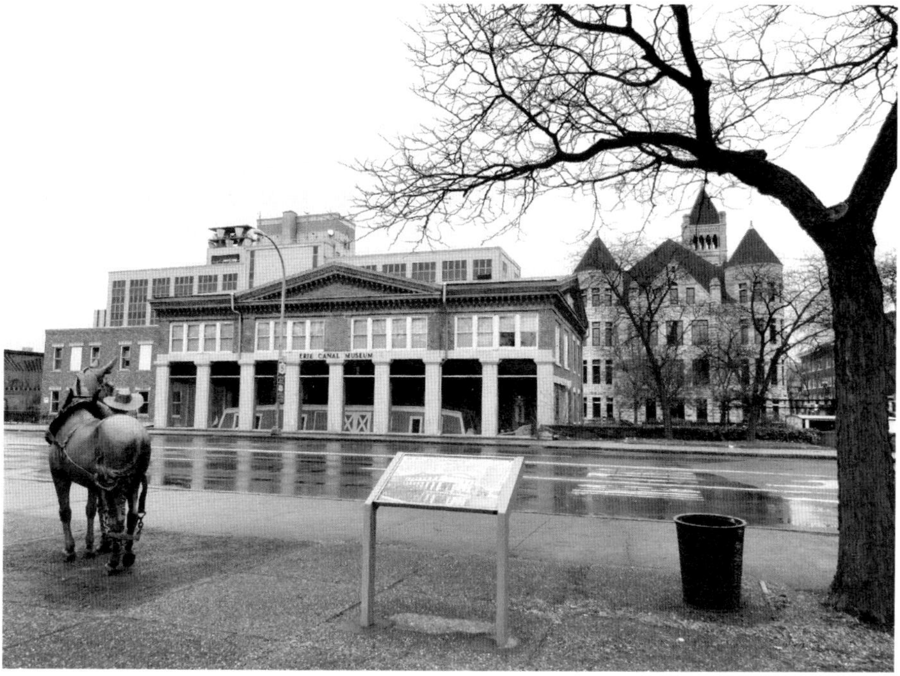

Erie Boulevard looking toward the Weighlock Building, which is now the Erie Canal Museum. *Neil K. MacMillan.*

action by witnesses. Sometimes, though, disaster doesn't kill outright. In the case of the Onondaga Creek culvert collapse, the repercussions echo to this day. The demise of the Erie Canal severely affected industry in the county. The brewing industry was severely affected by the collapse and would suffer even more harshly because of Prohibition. The last locally brewed beer in Syracuse—Congress beer by Haberle brewery—sent its last bottle into the world in 1962. Local beer not made by a mass-production brewery would not return until the advent of micro-breweries like Empire brewing and Middle Ages Brewery, both based in Syracuse. A way of life also vanished when the Erie Canal closed in Syracuse. Vestiges can be seen at the Erie Canal Museum in Syracuse and the Erie Canal Park in Camillus. One day in July 1907 changed Syracuse and Onondaga County forever. That is the legacy of the day the bottom fell out.[41]

12
TROLLEY FOLLIES

There was a time when trollies were the fastest and safest form of public transportation. Much as Centro® buses navigate Onondaga County roadways today, trolleys got the county residents from point A to point B in a mostly safe manner. The first trolleys were drawn by horses and ran on tracks, as did their steam-driven progeny. Trolley companies figured in political scandal and in the growth of Syracuse during the Industrial Revolution. Just as bus lines meet a need for those who are not affluent enough to afford an automobile or those who don't want the hassles associated with driving in urban areas like Syracuse, trollies filled that void in the waning days of the nineteenth century and the early days of the twentieth. Like their larger cousins, the railroad trains, trolleys ran down the streets of Syracuse and through the center of towns throughout Onondaga County. Advances in technology—notably the automobile and development of the bus coach as a form of transportation—spelled the demise of the trolley as a form of public transportation.

In an accident that shouldn't have happened, Michael Reiley was killed when the omnibus he was riding on collided with another omnibus. The Salina bus collided with a bus from the Cicero line. Reiley was the only fatality, but several passengers were injured when the body of the bus separated from the undercarriage. The Salina bus was trying to pass a manure wagon when it collided with the wagon. It was rumored that the Salina bus and a Cicero bus were racing when the accident occurred. Reiley's death was ruled to be a result of reckless driving on the part of the

Cicero driver and a failure by the Salina driver to exercise greater caution in avoiding a collision.[42]

A trolley car traveling on North Salina Street overturned after it left the track on June 3, 1895. The car rolled down a twenty-foot embankment. Twenty passengers were aboard the car at the time. All of them suffered minor injuries.

April 21, 1901, would not have been an auspicious day to ride the trolley. Mrs. James Forsyth was hurrying to catch a trolley with her head down when she was struck by the car. Weather conditions were stormy, and it is possible she did not hear the car or realize how close she was because of the inclement weather. Mrs. Forsyth was too severely injured to be taken to her home on Gifford Street and refused to go to the hospital. She was treated at the residence of the postmaster, who lived across the street on Burnet Avenue where the accident occurred. The victim suffered several lacerations. Wheeler Smith, a conductor on the Solvay line, fell off his trolley after being struck by a horse on the same day on Marcellus Street. He was treated for his injuries at St. Joseph's Hospital. He had been on the running board of the trolley car collecting fares when the incident occurred.

The Lakeside Railroad handled thousands of passengers annually in 1901. The trolley line was equipped with double-decker trolley cars. There had been accidents previously, and the Lakeside Railroad was perceived as one of the safer lines operating in Onondaga County. On June 5, 1901, a heavily populated car of the line jumped the track and tipped over. Twenty-three passengers, including five players from the Syracuse Stars baseball club, were injured. The motorman stated that he tried to apply both the hand brake and the electric brake to slow down, but both failed. The newspapers of the day wasted no time in castigating the management of the Lakeside Railroad. The accident happened at the corner of West Belden Avenue and Liberty Street. Investigators determined the track and wheels were not the cause of the accident; the car was traveling at a speed that would not allow it to negotiate the curve where it overturned safely. The motorman was attempting to slow the trolley down, and it was deemed that he did all he could to avoid the accident. The Lakeside Railroad stated that prior to the accident, the car had been rigorously inspected and both the power brake and hand brake were in excellent condition and fully functional. The Syracuse Stars played their game the night of the sixth in spite of having five players suddenly on the disabled list. Syracuse newspapers called for greater regulation of the streetcars.

A Syracuse woman, Mrs. B.W. Soper, was severely injured when her carriage was struck by a trolley on Wilbur Avenue. Mrs. Thompson and her two-year-old daughter, who were passengers in the carriage, suffered minor injuries. Although Mrs. Soper was not expected to recover, she regained her health.

We tend to think of automobiles trying to beat the train when crossing tracks as a modern phenomenon. It isn't—nor is it limited to automobiles. On August 19, 1910, Mr. and Mrs. Elmer Hayden were driving in their carriage on Lakeshore Road. Mr. Hayden saw the Syracuse to South Bay trolley approaching but thought he had sufficient time to clear the crossing. He was wrong. The trolley motorman, William Dewey, applied the brakes but was unable to stop the trolley in time. Subsequent investigation cleared the motorman of any culpability. It was determined that Dewey's quick response kept the accident from being much worse than it could have been. The Haydens were treated for their injuries. Those tracks just are not as clear as you think they are.

Jessie Kneeland was struck by a Syracuse and Auburn trolley on September 22, 1913, as she attempted to cross the trolley tracks near Dando's Crossing in Skaneateles. The fifty-five-year-old Kneeland believed the trolley was a local one and would stop at the station. As she stepped over the outside rail, she was struck and thrown into the air. The trolley stopped, and the injured woman was taken to the office of Doctor G.E. Clark in the village, where she was pronounced dead.

James Morris was riding his bicycle in June 1914 when he was struck by a trolley. He sued the trolley company for $500,000 for injuries sustained in the accident. Morris stated he was riding in between the track and the curb going the same direction as the trolley and crossed the track when he tried to avoid a delivery wagon ahead of him. The fore wheel of the trolley car ran over his leg, resulting in the loss of the leg. Morris alleged the trolley gave no warning before he was struck.

Michele Mollick sued the New York Railway Company in November, alleging the trolley he was riding went off the track, causing him to fall from his seat to the floor and suffer a rupture. Ironically, the infant Mollick was holding at the time of the incident suffered no injuries at all. He asked for $5,000 in damages. The incident happened near Assumption Cemetery on Court Street, and Mollick filed suit in November 1914. Given the fact that the child Mollick was holding suffered no injuries, one has to wonder about Mollick's claims.

Frank J. O'Neil died on October 13, 1915, when his head struck a tree. The New York State railway employee leaned out of his Liverpool line trolley and was struck by the tree. The county coroner determined that his death was accidental and that further inquiry into the causes of his death was not needed.

The blizzard of 1925 didn't just affect automobiles. The storm that blasted Onondaga County with copious amounts of snow also left slippery roads and trolley tracks in its wake. By Saturday, January 3, the county was digging out and life returning to normalcy. The New York Railways Solvay line car No. 1026 made its way on South Wilbur Avenue on its regular Saturday morning run. At 10:45 a.m., the rear wheels of the car left the tracks. The motorman valiantly fought to stop the car and regain control, but the slick, icy tracks and street made his endeavors futile. Car No. 1026 slammed into an elm tree at the corner of Gifford and Magnolia Streets. Fifteen passengers were injured in the crash. As the trolley left the tracks, a second trolley passing the H.H. Franklin works stopped as soon as it could to render assistance to the stricken Solvay line car. The incident was the second in two days; a car on the Summit line also ran away but was brought under control and no injuries were suffered. The trolley also struck a roadster, slightly injuring the two occupants, who proceeded to help the injured on the trolley once they extricated themselves from their stricken vehicle. During the ensuing investigation, the motorman, Andrew O'Donoghue, stated that neither the air nor hand brakes operated. The brakes were not on when police investigators examined the trolley wreckage, but investigators could not say the brakes were not on when the trolley jumped the tracks.

Sixty-year-old Edward Derusha of Syracuse died after a South Bay Line trolley struck him. Derusha was attempting to flag down the trolley at the Wolf Street stop when he was fatally injured. He was taken to Saint Joseph's Hospital, where he died of his injuries on the morning of January 15, 1929.

The days of the trolley have faded into the mists of time to be remembered vaguely behind the veil of rose-colored glasses when high school students or local touring companies perform the musical *Meet Me in St. Louis* or those who grew up in the waning days of the streetcar tell stories of shopping in downtown Syracuse and riding the trolley home. There are those memories, but there is also the reality of the Gifford Street accident. The trolley might be gone, but we need only look at the Megabus collision to see that the days of the streetcar still have lessons to teach us in the twenty-first century. Time will tell if we heed them in our ever connected world. Failure to will lead to more incidents like the Gifford Street accident and all the other trolley follies of our past.

13

OUT GO THE LIGHTS!

The Blackouts

Blackouts are usually more of a nuisance than a disaster, but sometimes they become both. The further we get into the twenty-first century, the less frequent power outages become. While blackouts occur due to natural reasons such as high winds, I focus here on those caused by equipment or human failure. Usually, natural failures of the power grid are extremely localized.[43] Early on in the electrical era, massive blackouts were unheard of. There were no interconnected systems even in the mid-1940s. If it is true that computerization made the world smaller, telecommunications gave the first push in that direction. In our increasingly connected world, we are able to communicate faster, but we are also more vulnerable to massive infrastructure failures. Not only do we stand to see failures caused by nature but also by sabotage, equipment malfunction, human error and computer viruses. These days, personal generators are available, but most run on gasoline. If you can't pump the gas if you run out, your generator makes a fine pigeon roost. We'll revisit these topics momentarily.

It was a quiet Tuesday evening. In Onondaga County, the evening rush hour had started as workers turned out from offices and factories. By 5:22 p.m., rush hour was well under way when the power winked off. The great Northeast Blackout of 1965 reached Onondaga County.

The blackout started in Ontario, Canada, when a 230-kilovolt transmission line tripped, triggering a cascading failure that plunged all or parts of eight states and the provinces of Ontario and Quebec into darkness.[44] The blackout was caused by human error. Scant days earlier, maintenance workers set a relay for testing purposes and neglected to reset it. When the power line failed, the improperly set relay allowed the cascade. In Onondaga County, the effects were annoying at first. County residents took the blackout in stride. One pedestrian in downtown Syracuse noted, "Well there's a full moon so I guess it won't matter if the power is of all night."[45] Several minor fires were reported in the county, likely the result of improper usage of candles or kerosene lanterns, but no fatalities were reported. Syracuse also noted that no muggings were called in. A freshman student at Syracuse University was transported to Upstate Medical Center with a sprained ankle. Traffic moved slowly due to traffic lights being inoperative, but by 6:00 p.m. the streets were cleared enough for traffic to move smoothly.

One might be forgiven for thinking the blackout doesn't really count as a disaster. Let's face it: there were no massive casualties or property damage, but economically almost every business in the county was forced to shut down. For industries like General Electric, Carrier, Fisher Body and New Venture Gear, the loss of two shifts' worth of product had an effect on their bottom line. For smaller businesses, the loss of a day's take could mean the difference between making a profit and hovering in the red. Hancock Airport was closed to all except emergency flights due to the runway lights being inoperative. Military communications were not affected by the power outage.

Almost as soon as the lights went out, rumors started about the cause. One caller asked the Syracuse Police if the blackout was due to an act of war. He was assured it was not.[46] President Lyndon Johnson was apprised of the situation at the Texas White House and was quickly assured that the power outage was not the result of sabotage.

Some people speculated later that the blackout was caused by UFOs. They allege the blackout occurred after a large, glowing red oval was spotted in the air over power lines in Clay, New York.[47] Another object, perhaps the same one, reported as a fireball was seen near Camillus, New York. Among the witnesses reporting these phenomena were the pilot and passenger of a private plane and the assistant aviation commissioner for the City of Syracuse.[48] Looking back, the causes for the Great Northeast Blackout of 1965 are clear, and the sightings of extraterrestrial objects do not factor into them. Human error led to a failure that quickly got out of control.

Was the blackout of 1965 a disaster? Yes, to a small degree economically. It was also a harbinger of things to come and scenarios that could yet happen. Thirty million people were in the dark for as much as twelve hours. It could have been so much worse! The blackout happened on a day that was warmer than usual for Onondaga County. At the time, many if not all furnaces were controlled electrically, just as they are today. The temperature was in the mid fifty-degree range. Sweaters and blankets were enough to keep the chill at bay, but a ten-degree drop in temperature or blizzard conditions could have proven catastrophic.

I like to tell people to err is human, to really mess up takes a computer. While I wish I could claim to be the originator of that statement, honesty must prevail. By 2003, computers were controlling many functions in power distribution systems. Computers have made switching power to less used lines and the relay of information much quicker in the half century that has passed since the blackout of 1965. Thirty-eight years made a difference in the power grid for not only Onondaga County, but the nation as a whole. Computers should make spotting problems and averting them quicker. Interconnectivity on the grid makes for better transmission and many times cheaper transmission of power. It also allows for surges to be compensated for—or it's supposed to.

Thursday, August 14, 2003, was a balmy summer day in Onondaga County. As rush hour approached, children were enjoying their last weeks of summer vacation, workers on the day shift were heading home looking forward to the upcoming weekend and those on the night shift were starting their daily grind. At approximately 4:11 p.m. the lights snapped off. The second biggest blackout in history to that point began. There was no spectacular cause. We couldn't blame it on Al Qaeda. The explanation was much more mundane. Power lines in Ohio failed when they brushed against branches of a nearby tree. The failure should have been localized, and indeed it would have been, had there not been a bug in the power company's computer system that kept an alarm from sounding in the control room for over an hour. Had the alarm sounded when the lines failed, operators in the control center could have isolated the problem and kept the power outage localized to Northeastern Ohio. Instead, the failure cascaded out of control. Fifty-five million people in eight states and the province of Ontario lost power.

All of New York State was in the dark, except a few pockets in eastern Long Island that relied on local power stations and were not tied into the grid; the area around Latham, New York; and areas in the southern tier that were hooked into Pennsylvania's grid. In Onondaga County, there were no fatalities, and power was restored around midnight. Businesses shut down or severely curtailed operations during the blackout. Nuclear plants, including the reactors at Nine Mile Point Oswego County, throughout the affected region shut down automatically as a safety precaution. The City of Syracuse stepped up police patrols to prevent looting, and law enforcement agencies stepped up operations to counter the negative effects of the blackout. There was no looting. Again, in the wake of this blackout—as in the 1965 power outage—the economy suffered from the shutdown.

Was it a disaster? Perhaps not, but gas prices rose and locally, restauranteurs not only lost business but also had to discard food due to spoilage. Almost as soon as the power blinked out, the age old blame game started. Canadian officials said the problem was with New York. Governor Pataki asserted it was a Canadian issue. Both were wrong. The initial cause was the tree branch and the failing alarm. The blackout of 2003 laid bare the fundamental flaws of our aging infrastructure. A telephonic warning system put in place should have forewarned most of the agencies affected by the power outage. It didn't work; someone forgot to pick up the phone. A simple phone call could have spared many of the areas the outage. Again, Onondaga County dodged a climatic bullet. The blackout occurred on a summer day when the temperatures were in the upper eighties. Had the blackout happened in mid-January instead of mid-August, the results could have been dramatically worse.

What are the lessons of these two blackouts? Better warning systems with better fail-safes are needed as we move further into the twenty-first century. At any point, the power grid is vulnerable to not only nature and human error—as proven by the blackout of 2003—but also to sabotage by those who would harm this country. No matter how much we engineer a better system, we still are human. In 2003, power company officials in New York State, including those from Niagara-Mohawk, stated that there were no failures in the state's grid. They were right. Had there been adequate warning, the state's power authorities could have unhooked from the system, and the blackout would have not affected so many people. Experts had for years before the 2003 blackout warned that the system was due for a major crash. They are still stating that. In the days since the 9/11 attacks, we worry about terrorists mounting an operation that kills multitudes. It is a valid worry, but so is the idea that they might launch a strike to cripple our infrastructure through overt action such as bombing a relay station or a stealthy attack by hacking the system. The more we are connected, the more widespread the effects of such a strike.

14

SLIP SLIDING AWAY

The Tully Mudslide

The year 1993 may well go down in history as the year Mother Nature showed Onondaga County who was boss. A late winter blizzard paralyzed the county in March, and April was one of the rainiest on record. In between snow melt and rain, many areas suffered minor flooding, and the ground was saturated enough for farmers to worry about getting crops planted in time. Onondaga County is not known for mudslides. For most of us in the area, mudslides were something we saw on television in California. The hamlet of Tully in Southern Onondaga County is a largely rural farming community. By the last week in April 1993, farmers were looking forward to getting their crops put in.

Generally, the soil in Onondaga County is stable. The soil on Bear Mountain along Tully Farms Road had been stripped of tree cover and plowed in preparation for sowing the season's crops, and a record snowfall in the winter and heavy rain saturated the soil. Water filtered through the topsoil and collected in the layer of gravel and sand beneath. The clay layer below the gravel acted as a waterproof barrier. A tour bus of scientists and mining experts transited Tully Farms Road as part of a geology lecture tour to view mud boils at 12:15 p.m.[49] The experts were luckier than they knew. Forty-two minutes later, a fifteen-foot-wall of soil and rock careened down the hill. The tidal wave of mud destroyed one house, damaged two others and buried a pick-up truck while totally blocking 1,400 feet of the two-lane road. Experts quickly ruled out the mud boils. Calls went to 911 as soon as the mudslide started. Helicopter rescue crews plucked three people from the mud. The Town of LaFayette, which governs the hamlet of Tully, quickly declared a state of emergency and sealed off the affected area. Fortunately, none of the people living in the area was injured. A Belgian sheepdog was killed in the mudslide, and several others were missing.

Three families were displaced. To add insult to injury, their insurance didn't cover the mudslide. The state and federal government quickly stepped in. The federal government promised funds to help offset the cost of cleaning up the fifteen-foot-high mud wall. The National Guard was called in to seal off the area from gawkers. Several were caught as they tried to access the area via old logging roads. A minor mudslide shut down the southbound lanes of Interstate 81 for a while two days after the Tully mudslide. In that instance, no mud clogged the highway, but DOT officials closed Route 81 as a cautionary measure. Officials estimated cleaning up Tully Farms Road would cost between $100,000 and $115,000. The federal government spent $138,000 on the cleanup. June Horton's home was valued at $100,000, and she raised her family there. The home was destroyed by the mudslide. Standard homeowner insurance policies in New York State specifically excluded damage by mudslide from coverage. Ironically, though the mudslide knocked the house off its foundation and punched holes through the cinder bricks, not a picture fell from the walls. Jim Fylers's house was destroyed in the cataclysm. Linda Elligott endured the loss of one of her Belgian sheepdogs when the wall of mud crushed her kennels. Three others were rescued; two others were missing and presumed dead. Elaine Van Dee's 170-year-old house was spared, but it was a close thing.

As devastating as the mudslide was, had it struck two hours later, school buses would have been transiting the road and discharging their passengers. Seven homes were damaged in the mudslide. Soil was put on farmer's fields in the area or removed temporarily to the site of the LaFayette Apple Festival. State senator John DeFrancisco introduced legislation to render aid to the victims of the mudslide and flooding that occurred in Onondaga County.

What caused the mudslide is an easy question to answer. In drier years, the clearing and plowing of that section of Bear Mountain would not have caused a problem. The spring of 1993 followed a winter that saw a record 192 inches, or 16 feet, of snow in the area. In most years, Onondaga County sees approximately 120 inches of snow, and it is interspaced throughout the winter with warm spells that allow it to melt without the build-up that was seen in the wake of the March 13 blizzard. The month of April was rainier than usual. While the northern portion of the county saw localized flooding, Bear Mountain soaked up water until something had to give. When the field

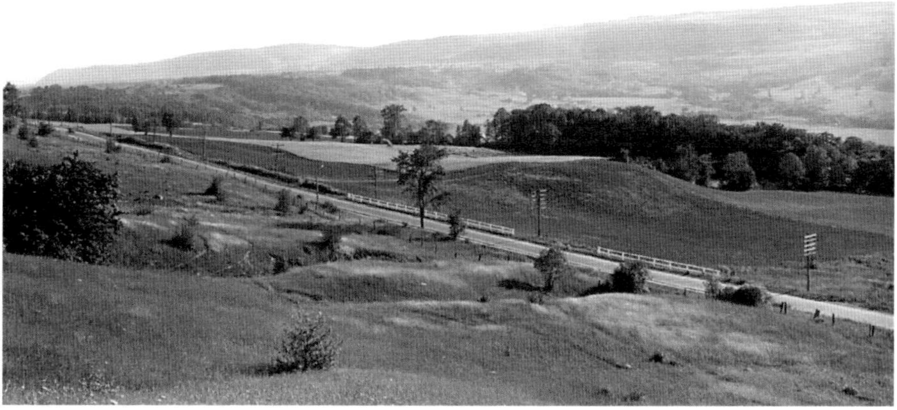

The hills of southern Onondaga County. A hill similar to these suffered a mudslide in Tully after an excessively wet spring. *Courtesy of the Library of Congress.*

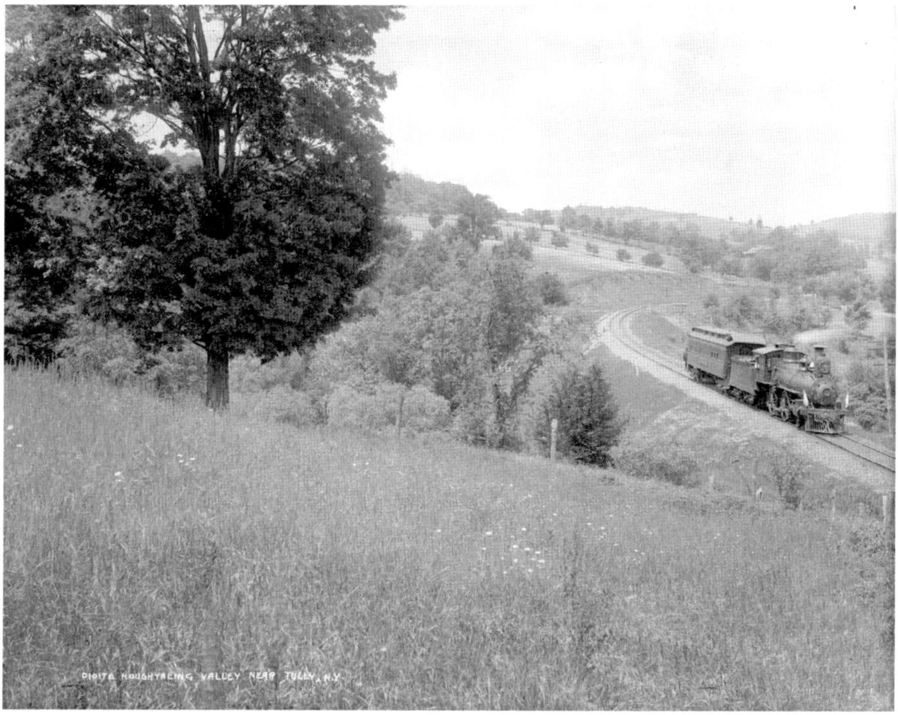

Houghtaling Valley, near Tully. A mudslide near here in 1993 destroyed four houses but fortunately killed no one. *Courtesy of the Library of Congress.*

on Bear Mountain was cleared and plowed, there was no established root structure from trees to hold the soil in place. It would be easy to point a finger and say, "You shouldn't have plowed," but 1993 was an unusual year for weather in Onondaga County. Residents of Tully Farm Road stated they expected a mudslide at the southern end of the valley, where mud boils were prevalent. The mud boils usually add sediment to Onondaga Creek.

Response time and efforts cannot be faulted. Within minutes, first responders were on the scene. The devastation of the mudslide was shocking enough. The insurance companies' response to the afflicted came as just as much of a blow. Homeowners were not told at the time that their policies did not cover mudslides. In Onondaga County, mudslides aren't something that occur frequently enough to warrant asking if they are covered. Legally, the insurance companies were not obligated to cover mudslides in a standard homeowner's policy. As I stated above, like earthquakes, mudslides are not a common occurrence in Central New York. An unusual set of circumstances led to the Bear Mountain mudslide of 1993. Hopefully, Onondaga County will be spared a repeat, but it is a possibility. Record snow followed by heavy rain and a dearth of a viable root system led to the mudslide. Tully remains a beautiful, pastoral setting to build or make a home, but the hamlet is still vulnerable to the whims of Mother Nature.

15
ENVIRONMENTAL CONCERNS

A Dike Break, an Acid Spill and the Lake

There was a time when Onondaga Lake whitefish were a delicacy eagerly consumed by the ultra-rich in the city of New York and points beyond. The lake was a sportsman's paradise. Even earlier, the Onondaga people built their villages along the shores of Onondaga Lake, and the Haudenosaunee Confederacy came into being near the village of Liverpool. The waterways in Onondaga County provided food and water and were the highways for both the Onondaga and the first Caucasians to see the area. As Onondaga County came into being and grew, we took the waterways for granted. Progress was the driving principle, and the region grew at the expense of the environment.

As the county grew, we collectively used our waterways as sewers. Early on, outhouses leached their contents into the soil, and from there, the contents made their way to our waterways. With the reintroduction of indoor plumbing, septic tanks held the sewage, where it broke down. As we shall see in the next chapter, this improved health in the county but still played havoc with the area's waterways. Sewage treatment helped alleviate the problem a bit, but with progress came industry. The factories that gave consumers the goods they desired and armed our nation in times of war generated copious amounts of waste products. In some cases, these byproducts were dumped directly into area creeks. For most of the county's history, there were no real laws to prohibit dumping into our waterways. Today, even with stringent environmental laws, enforcement can be haphazard due to budgetary constraints. It seems logical that you wouldn't want to put anything that would affect water quality into the very water we drink and bathe with. As I type this, there is a limit on how many fish you can eat from Onondaga Lake. There was a time when the waters feeding Onondaga Lake could safely be drunk without boiling or any other preventative measures. Sadly, those days have gone the way of the carrier pigeon. When did we forget that water is life?

Onondaga Lake has several tributaries. Among these are Nine Mile Creek, Bloody Brook, Onondaga Creek and Ley Creek. Over the years, Ley Creek has been the subject of several lawsuits by residents seeking to halt municipalities and industries dumping waste into the creek. In 1913, a farmer by the name of Mautz claimed that several industries were pouring waste and refuse into Ley Creek, polluting the waterway to the point of being unfit for use on his farm. Consolidated Rendering officials and other industry advocates claimed the water was treated before being pumped into Ley Creek. In 1914, Justice Williams, the judge trying the case, awarded Gilbert Mautz $400 and issued an injunction against the Village of East Syracuse to cease dumping sewage into the creek. The companies and the village appealed the decision. Williams's decision was upheld in 1918.

Ley Creek meanders for ten miles before flowing into Onondaga Lake. The sleepy little creek was formed when the last glaciers moved through Onondaga County ten thousand years ago. The creek has been referred to as a menace and during the 1930s acquired the moniker "Old Stinky" due to the years of sewage and industrial pollutants dumped into its waters. Ley Creek supplies Onondaga Lake with around 8 percent of its water. That doesn't seem like a lot, but it isn't the lake's only tributary. There is an old adage, "If you put a drop of sludge into wine, you have sludge. If you put a drop of wine into sludge, you have sludge." For over one hundred years, Ley Creek has been a ten-mile-long sewer. Ley Creek is also the third-largest tributary of the lake.

In 1926, the Eastwood sewage treatment plant was rebuilt. After several years of being in disrepair, the plant was pouring raw sewage into Ley Creek. Workers finished construction on a new sewage treatment facility on Seventh North Street in 1935. Almost as soon as the plant opened, it faced problems with overflows. Over nine million gallons of raw sewage were being shunted to the much maligned creek. While sewage is biodegradable, it feeds ammonia, phosphorus and bacteria into the water that robs the water of oxygen.[50] Flora and fauna in the creek suffered from the loss of oxygen.

Gilbert Mautz was prescient in suing, but he would have been shocked at later events. By the 1940s, a cornucopia of county industries were pouring

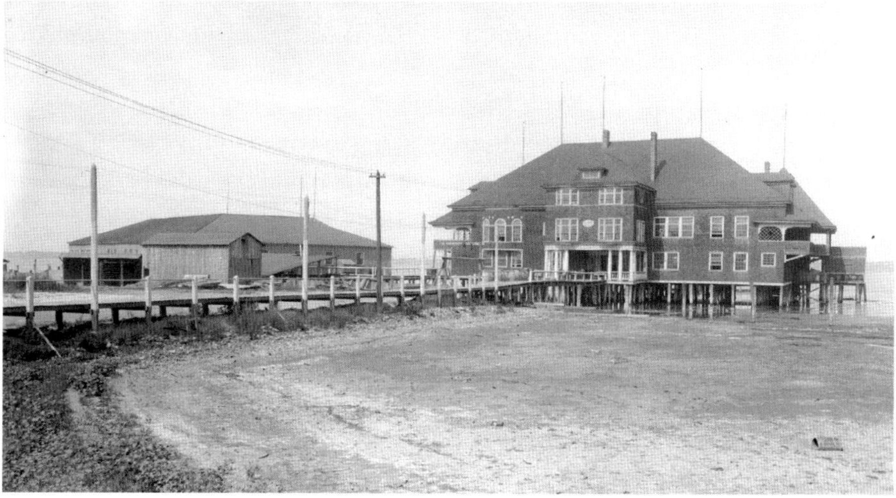

Onondaga Lake near the Onondaga Yacht Club in 1890. The lake was once the source for several species of game fish, and Onondaga lake whitefish were a sought-after delicacy in the finest restaurants.

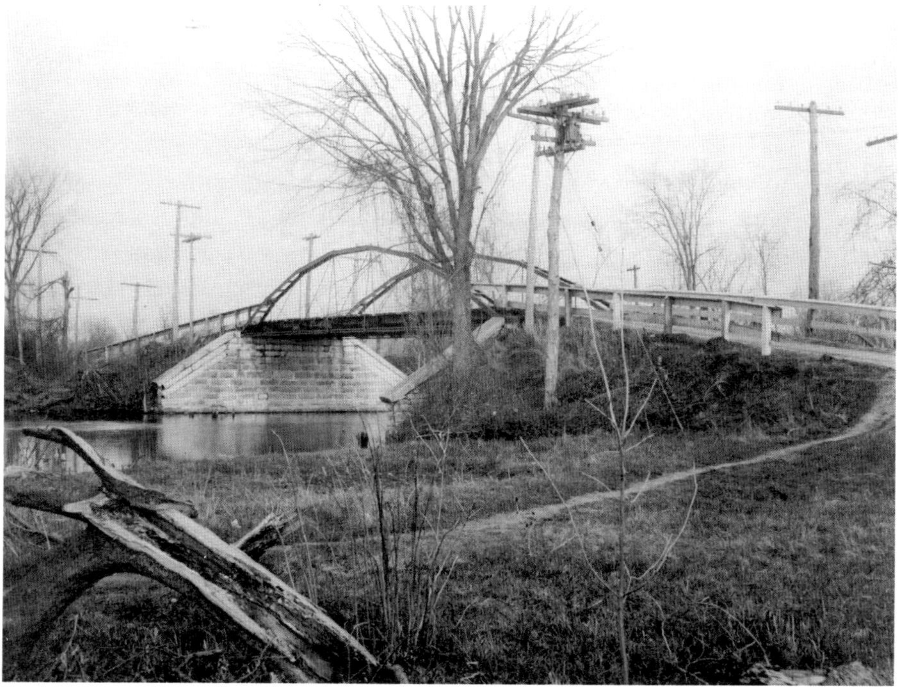

Onondaga Lake outlet around the time of the 1912 tornado. *Courtesy of the Library of Congress.*

lethal amounts of chemicals such as PCBs, ketone and heavy metals into Ley Creek. In 1973, the federal government passed the Clean Waters Act. The *Post-Standard* urged in an editorial that a "Sewage Czar" be appointed to oversee efforts to curb pollution entering the state waterways. At the time, Onondaga Lake was considered one of the most polluted lakes in the United States, and Ley Creek was a major contributor to the problem. In the 1980s, candle manufacturer Will and Baumer was cited for dumping phosphorus and other waste products into Ley Creek. Hancock Airport allowed thousands of gallons of aircraft de-icing fluid to drain into Ley Creek. By 1993, Ley Creek was considered one of New York State's top ten endangered waterways. All of that pollution was making its way into Onondaga Lake.

By 1905, Onondaga Lake whitefish were gone from the lake. For many years, they had been a delicacy that upscale restaurants paid top dollar for. Overfishing and the ravages of pollution took their toll. The lake was still a draw for tourists and residents alike at that point. Several amusement parks and resorts dotted the shore. By 1940, swimming had been banned in the lake. By 1970, there were strictures against eating any fish caught in the lake. Ley Creek wasn't the only culprit. Since the late nineteenth century Solvay Process Company had manufactured soda ash and other chemical products widely used throughout industry. Decades of disposing of the waste products in the lake aggravated the pollution. Among the products dumped were copious amounts of mercury. On Thanksgiving Day 1943, a retaining wall to keep a slurry of waste products broke. The waste blocked State Fair Boulevard and inundated the fairgrounds. Investigators quickly determined that sabotage was not the cause of the breach. Workers toiled to mitigate the damage. In subsequent years, Solvay changed hands until it was finally purchased by Honeywell. When the plant finally closed, Honeywell was tasked with cleaning up Onondaga Lake. Much of the pollutants in the lake were cleaned out by dredging, completed in 2014. The remainder was capped. The Onondaga Nation claims, with some justification, that capping is at best a temporary fix. Today, you can consume a limited amount of fish caught in Onondaga Lake. Onondaga Lake and its tributaries were designated a superfund site in 1994.

Let me be clear that industry was not the only culprit. For many years, raw sewage was dumped into the lake's tributaries. Even after the county established sewage treatment facilities, there were overflows when the system was overwhelmed. A 1989 agreement between Onondaga County and the EPA called for improvements to the sewage treatment system in an effort

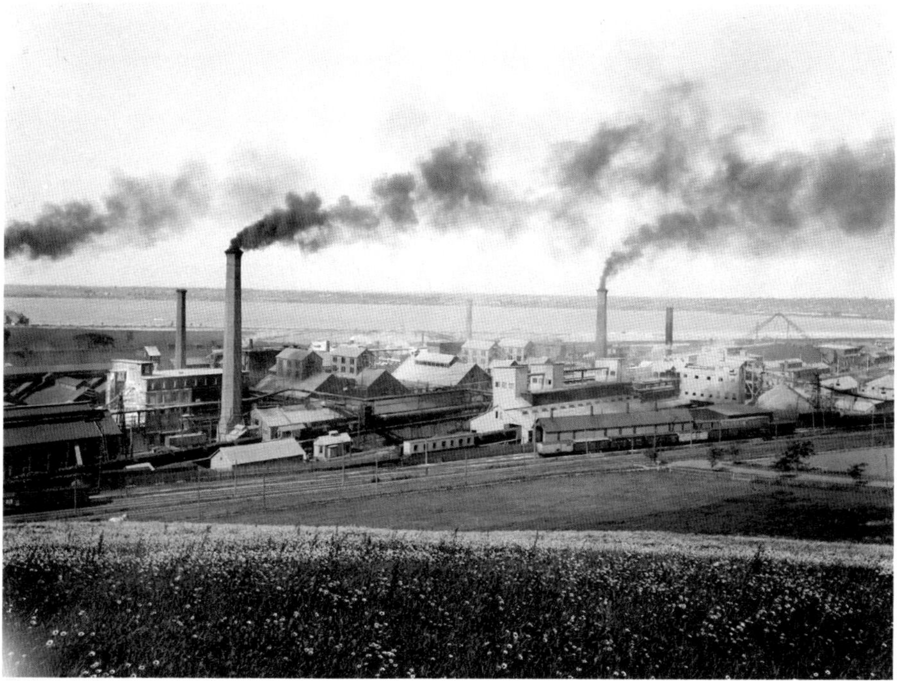

Solvay Process Company in the 1890s. The company was closely intertwined with the county's history and ecology. *Courtesy of the Library of Congress.*

to curb discharges of ammonia, phosphorus and overflow waste into the lake and its tributaries. By 2014, Onondaga County had abated forty-six of seventy-two combined sewage overflows. We are not immune to storms here in Onondaga County, and all that storm water runoff has to go somewhere. All the trash, salt from highway snow removal, fertilizers and pesticides that are sprayed on our lawns and gardens ends up in the water. Newer fertilizers with less phosphorus are being manufactured, vacuum trucks are being used to remove refuse from storm drains and skimmer vessels are deployed on the lake to catch the garbage that does get into the lake. There is an active recycling program in Onondaga County.

This is one of the disasters that can be reversed with a bit of foresight and thought. While many look to government to solve our ills, the choices we make also play a part in the health of Onondaga County's waterways. Is there really a need for your yard to look like a putting green at Pebble Hill? We castigate industry for dumping poisonous chemicals into the lake, but still there are those of us who pour their used motor oil down

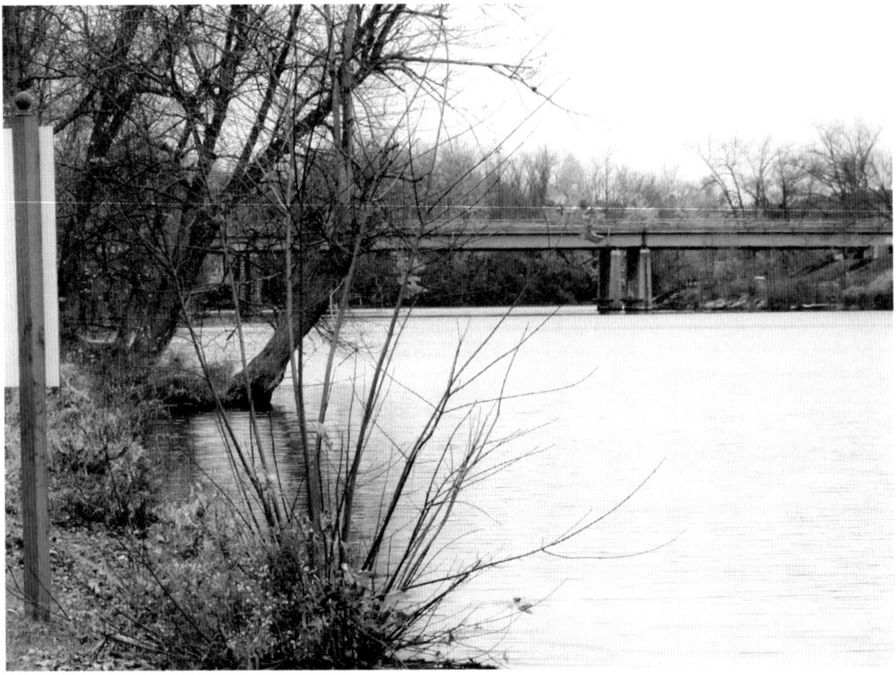

The Onondaga Lake outlet into the Seneca River today. Pollution in the creeks that feed the lake adds to the lake's environmental woes. *Neil K. MacMillan.*

the storm drains. We're still using rock salt on our driveways. There are still questions when it comes to cleaning up Onondaga Lake and its tributaries. The lake has improved in my lifetime, and with proper management, it will continue to improve. I don't know if we'll ever see the day when we can again eat the fish in the lake with no restrictions or see the ban on swimming lifted. Hopefully, the future will see a solution that truly cleans the lake so a cap is no longer needed. If the history of Onondaga Lake teaches us anything, it is that environmentally, all things are connected and garbage in equals garbage out.

16

AN ILL WIND BLOWS

Disease Runs Rampant

We don't think of epidemics as disasters. Hurricanes, tornadoes and earthquakes rank high on subjects for disaster movies, but disease doesn't. The previous three kill sometimes in the thousands. Epidemics can kill millions of people. The Black Death killed millions in Europe in the fourteenth century. When I was growing up, epidemics were something that happened in third-world nations. Today, we face the more than remote possibility of epidemics or pandemics striking within the next five years. I know that sounds alarmist, but bear in mind that the spring and summer of 2015 saw the Zika virus emerge from the Caribbean basin and the continuing emergence of antibiotic-resistant diseases. George Santayana said, "Those who cannot remember the past are doomed to repeat it." This is particularly true when it comes to disasters and epidemics. Onondaga County is not immune from the ravages of disease, and the county's history is littered with the corpses of epidemics past.

For generations, the word *cholera* struck fear into our ancestors. Cholera was first identified as a separate disease in the late eighteenth century.[51] North America was considered immune to cholera thanks to being separated by the Atlantic Ocean from Europe and, by extension, Asia. Even doctors in 1832 knew very little about bacteria and the mechanics of disease. The epidemic started in the United States when English immigrants infected with cholera disembarked in New York. Designed to get goods and people across New York State rapidly, the Erie Canal quickly became a river of death as people infected with cholera, not knowing they had the illness,

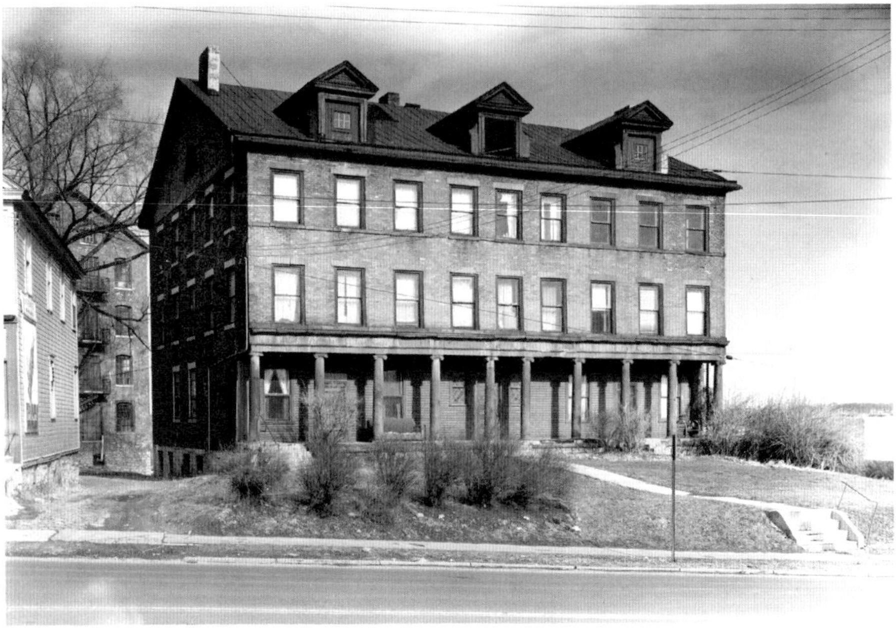

The Botanic Infirmary was one of Onondaga County's first hospitals, established to treat some of the area's first epidemics. *Courtesy of the Library of Congress.*

made their way through the Empire State. No community along the canal was spared. Buffalo, Utica and Cooperstown were all hit hard, but Syracuse and Salina suffered approximately 100 deaths due to the disease as well. In 1850, the population of Syracuse was 22,271. After years of immigration and population growth, 100 deaths would still have been a staggering blow to the community. The 1832 epidemic—so called in literature because that is when it struck New York State and New England—actually began in 1827 and was part of the second cholera pandemic, which lasted through 1835.[52] The second pandemic killed 150,000 in the United States. Cholera is prevalent in areas where sanitation and clean water are problematic. At this writing, cholera is listed as being pandemic. Cases in developed nations are rare, but in underdeveloped areas of Asia and Africa, the disease is a major health concern. Cholera is marked by muscle cramps, severe diarrhea and resulting severe dehydration.

If cholera frightened our ancestors, smallpox could be horrific to them. Smallpox is a viral disease that is spread through the air or by close contact. The disease is marked by a rash and blisters covering the body, and the survivors exhibit scarring as a result of the blisters. In 1874, a localized outbreak of smallpox struck Syracuse. In response, the City of Syracuse founded the City Hospital of Communicable Diseases. That year, 221 people died from the disease. By this time, there was a vaccination for the disease, and the government had vaccinated Native Americans for smallpox since the 1830s. Smallpox would scourge the world for another 104 years. The disease was declared eradicated in 1978. The World Health Organization (WHO) recommended destruction of all laboratory samples of the disease in 1986 and wanted an end date of 1993. Both Russia and the United States argued against the destruction on the grounds of scientific research. Some advocates state that the retention would allow new antiviral drugs to be developed and better understanding of viral diseases in general. Opponents, including the WHO, argue the existing batches could be used to reintroduce the disease in a weaponized form. For now, earth remains smallpox free.[53]

We get vaccinations for influenza yearly. Unlike smallpox, which has been eradicated and stayed constant, allowing the vaccination process to spell the demise of smallpox, influenza is a highly adaptable group of viruses. Vaccinations are usually ineffective after two or three years due the virus's ability to mutate. Several epidemics and pandemics struck the world and Onondaga County. Outbreaks were noted in 1847–48, 1850–51, 1857–59 and 1889–90, and while these were devastating, none would reach the ferocity of the 1918–19 Spanish flu outbreak.

America and the world was at war in 1918. As 1918 progressed and the killing continued, the Spanish flu struck Europe. Some scientists theorized the outbreak started in an army camp near Étaples, France, a contention that is still hotly debated. The disease swiftly spread across the globe. Unlike many diseases, the Spanish flu targeted the young and healthy. The name Spanish flu was coined in the mistaken belief that the disease originated there. In fact, the origin of the name comes from the fact that Spain, being one of the very few European countries not embroiled in World War I, was not censoring their newspapers as heavily as the combatant nations were and reported that their King Alphonso XIII suffered from the illness. In the close confines of army camps, the disease added rapidly to the ever-growing casualty lists. Nowhere on the globe was free from the disease, and cases were reported in the islands of Micronesia and in the Arctic. The Spanish flu resulted in anywhere from 50 to 150 million deaths, approximately 3 to 6 percent of the world's population at the time. Onondaga County was not spared the ravages of the disease. Then, as now, conspiracy theories rose in the wake of the plague. Dr. Arthur N. Davis, the former dentist to Kaiser Wilhelm the Second, theorized that German spies brought the disease to

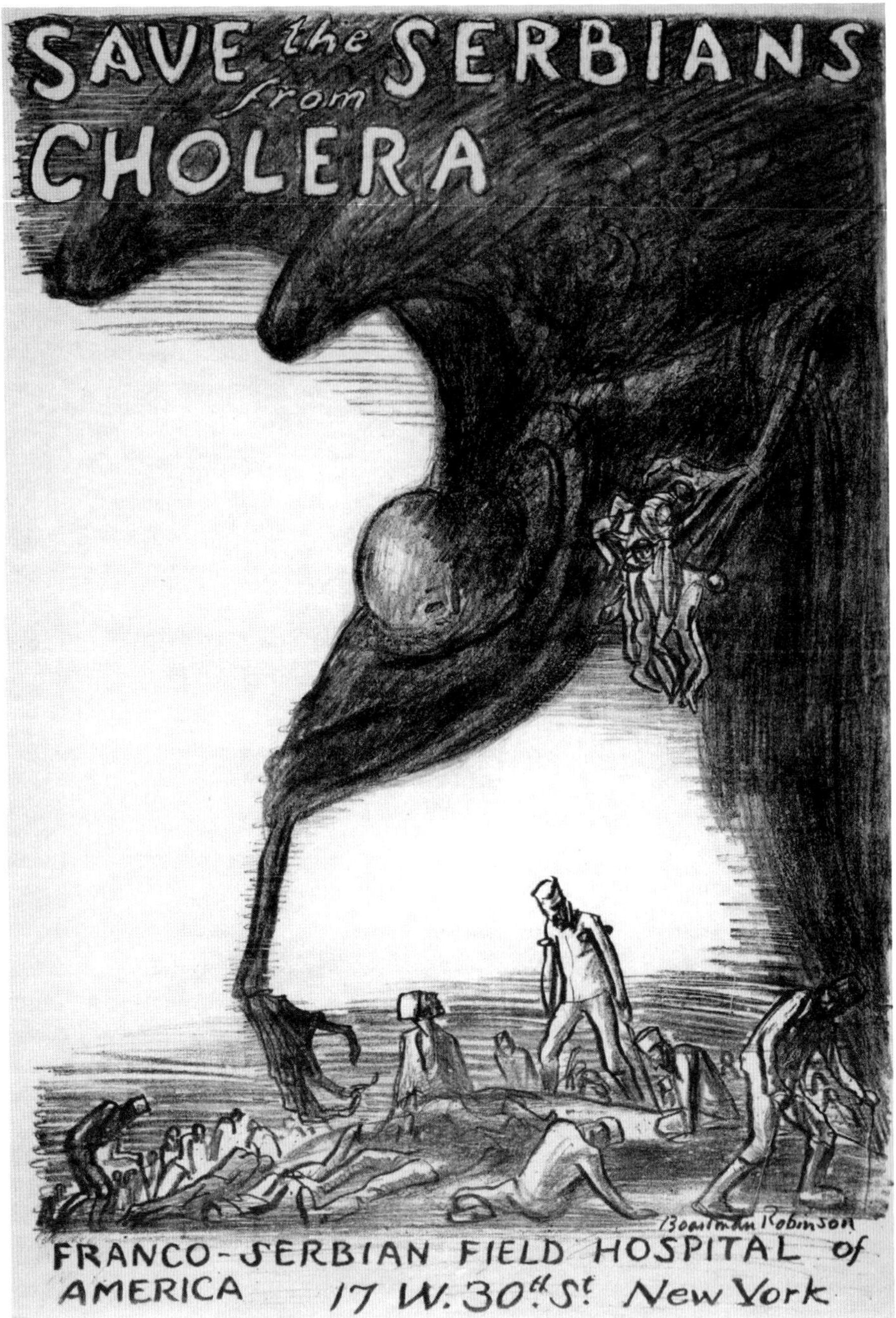

A 1918 poster requesting donations for the Franco-Serbian Cholera Hospital. Cholera was one of many diseases to ravage Onondaga County in the nineteenth century. *Courtesy of the Library of Congress.*

American shores. First reports of the Spanish flu were noted on September 27, 1918, in Onondaga County. It is certain that there were cases prior to that date, but New York State did not require reporting before that date. In Onondaga County, 12,000 residents were struck by the disease, and 253 people succumbed to the Spanish flu in Syracuse.[54] Ironically, for many years, the Spanish flu pandemic was all but forgotten. The disease was overwhelmed in memory by the 11 million deaths from the carnage of the Great War. The population was used to epidemics striking the world. The Russian flu had raged through the world a scant twenty years earlier, and an infantile paralysis epidemic had struck in 1916. In Onondaga County, the Spanish flu held sway for a short time compared to other places. By November, no more cases were being reported, and by the third week in October, restrictions on public gatherings had been lifted. Influenza still rears its head, and we still look for ways to combat the disease.

Infantile paralysis struck the county twice. The first epidemic stuck in 1916, a scant two years before the great Spanish flu pandemic of 1918. Poliomyelitis is caused by the polio virus and is transmitted through fecal matter and, less commonly, saliva. The frightening thing about infantile paralysis is the people the disease targets. The vast majority are young children. The disease can paralyze but doesn't always do so. Some victims show no symptoms. The disease develops rapidly. One case in August 1916 killed the victim in just two days. In Syracuse, patients stricken with polio were taken to the City Hospital for Communicable Diseases. Medical personnel put forth their best efforts to combat the disease, but all they could do was isolate the stricken to prevent infection and make them as comfortable as possible and hopefully use physical therapy to return the patient to a semblance of normalcy. In the 1916 epidemic, physicians utilized blood serum in an effort to counter the effects of the disease. The first cases in Onondaga County were noted on August 26, 1916. The polio epidemic had been raging earlier in the city of New York. By August 3, 1916, Dr. F.W. Sears, the Syracuse Health Officer, had noted that while Syracuse had yet to face an epidemic of poliomyelitis, he believed it was coming.[55] He was right. Three cases of death by polio reached the morgue, and seven other cases were being monitored. Among the recommendations made by Dr. Sears were to notify the health bureau of visitors from Philadelphia or New York. Experiments were being conducted with a compound made from the adrenal glands of sheep in efforts to curb the effects of the outbreak. Fatalities were reaching record limits by August 1, as Syracuse and Onondaga County had yet to be scourged. The first victim to recover from polio in the 1916 epidemic in Syracuse was released from the hospital on September 6. Schools opened on October 9, 1916.

In the late 1940s, poliomyelitis ravaged the nation again. In the interim between the 1916 epidemic and the latest outbreak, medical researchers made great strides in the fight against the disease. In 1952, gamma-globulin was the treatment of choice. Three years later, in April 1955, Jonas Salk's vaccine was available for mass immunization. As baby boomers received vaccines, incidents of polio diminished. Infantile paralysis is not eradicated. The disease can still be found in a handful of nations. The disease is almost unheard of in Onondaga County. A strenuous vaccine program and improving sanitary conditions in the intervening fifty-odd years has led to generations who did not have to deal with the scourge of polio.

AIDS—the very name scares us. Even cancer does not generate the fear that AIDS does. It is a disease fraught with misinformation and bigotry. In the early 1980s, in the wake of the sexual revolution of the late 1960s and early 1970s, a new illness struck America and Onondaga County. For cancer,

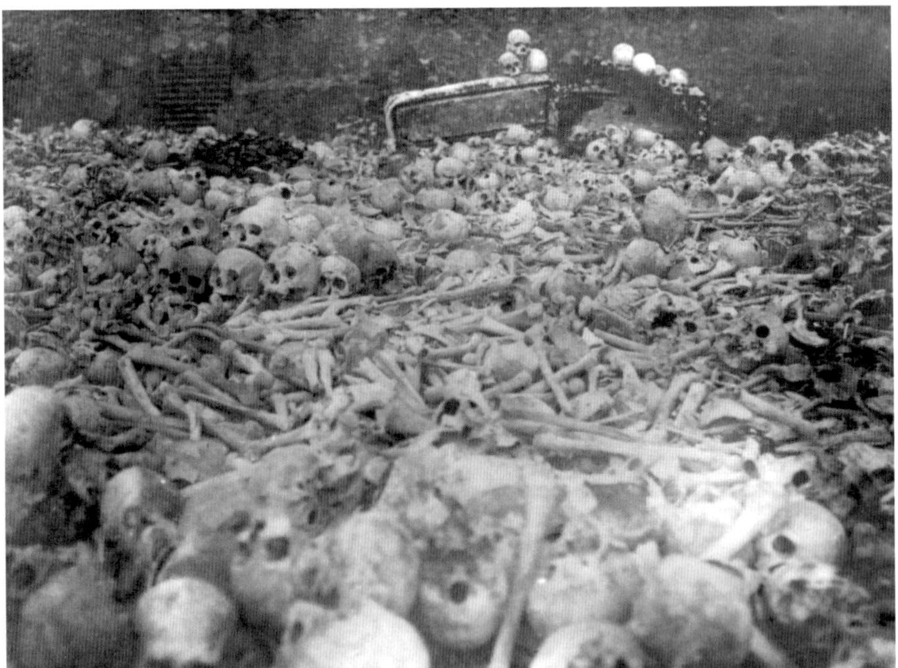

A pile of human remains in the Philippines in the early 1900s. Epidemic victims were sometimes buried in mass graves to prevent infection. *Courtesy of the Library of Congress.*

or most forms of cancer, there is no cure or vaccine yet. Unlike cancer, AIDS was seen as a disease that struck the fringes of society. Initially, it was referred to as the "gay disease" and associated with homosexuals and drug users. As the disease made inroads to mainstream society, the blame game started—not out of malice, but rather out of ignorance. Almost as quickly as cases started showing up, conspiracy theories about the disease's origin became endemic as well. The disease is believed, officially, to be the result of a virus found in green monkeys and transmitted to humans who processed bush meat.[56] However, others posit the theory that the virus was created in bioweapons labs and surreptitiously infected homosexual men with it under the guise of testing a new vaccine against hepatitis-B.[57] Notably, the USSR floated the theory during the Cold War, and it still rears its head occasionally.

In Onondaga County, cases climbed as the last two decades of the 1900s moved forward. In September 1989, Onondaga County reported 86 cases of AIDS resulting in 61 fatalities dating from 1982 to 1989. By March 1991, the number rose to 153, with 104 fatalities. By 2007, the number of AIDS cases had risen to 914, and county health officials were noting an increase in cases of young men coming down with the disease. Treatment improved in twenty-five years, and that was part of the problem. Some people believed they could get a pill, and all would be well. There was no cure for AIDS then, and there still isn't nine years later. Medical science has yet to find a vaccine for the HIV virus that causes AIDS. People are living longer, but the disease is still incurable. Prevention efforts have improved, and as we move further into the twenty-first century, much of the stigma associated with HIV/AIDS has disappeared. The measures for prevention are simple and have been touted for years. Get tested, use condoms to prevent exposure, avoid intravenous drug use and know your partner. Rates among drug users are above those of nonusers. As we move further into the century, we will improve how we treat the disease and hopefully find a cure. Until then, the disease will continue to be pandemic.

Those of us who grew up in the 1960s and 1970s remember seeing bumper stickers promoting "Sex, Drugs and Rock and Roll." In rebellion against what was seen as repression by adults, drug experimentation became, if not widespread, then commonplace in the news. Some of the greatest musical icons of the era, such as Janis Joplin, died from drug overdoses. During the late 1960s, heroin was the bogeyman we grew up hearing about. Stories abounded in the news and in rumor of robberies and murders committed so junkies could get their fix of heroin. For many years, heroin addiction was seen as an inner-city problem. Heroin, also known as diamorphine, was synthesized in Germany by chemist Felix Hoffmann, who was working for Bayer pharmaceuticals. The intent was to create a form of codeine that would be less addictive than morphine. All three are opiates. The result was a drug with one and a half times the addictiveness of morphine. This wasn't known at the time, and Bayer marketed diamorphine as a cough suppressant and treatment for morphine addiction under the trade name of heroin. In 1914, the Harrison Act outlawed the sale and use of heroin in the United States. By 1930, the League of Nations had followed suit. Diamorphine is still available in some countries with a prescription. In the United States, it is illegal to possess heroin without a license from the DEA.

Heroin seemed to disappear from the news in the last quarter of the twentieth century. Drugs like cocaine and later ecstasy became focal points for the press. Like a viper hiding in a wood pile, heroin lurked in the shadows. As the second decade of the twenty-first century opened, the viper bared its fangs. Crouse Hospital admitted heroin addicts in 2000. In 2010, there was 1 death attributed to heroin. By 2015, the number was a staggering 890. The drug is no longer confined to dark back alleys or seedy, run-down drug houses. More cases are coming out of suburbia every day. In 2006, the New York State legislature approved the use of Narcan by non-medical personnel to counteract overdoses. The drug does save lives and has been used with great effect. The *Post-Standard* for February 14, 2016, reported that Central New York was ill-equipped to handle the growing epidemic. It remains easier for addicts to obtain heroin than it is to get the methadone and counseling they need to help them combat their addiction. Many addicts start on the road to heroin these days because they are treated for pain with opioids and develop an addiction to the painkillers. When they can no longer get them through the doctor, they turn to heroin, which is a cheaper alternative. Like AIDS, the opioid epidemic is being combatted with education. Political rhetoric and the traditional stigma associated with drug addiction still stand in the way. The latter is decreasing thanks to educational efforts. Criminalization for users has thus far failed to curb the epidemic—so has throwing vast amounts of money without having a clear and concise plan to help those addicted. We are still looking for answers to quell this current epidemic in Onondaga County.

It seems certain that the future will hold more epidemics in store for Onondaga County. This summer saw the emergence of the Zika virus as a threat in the southern United States and throughout the tropical regions of the western hemisphere. Time will tell if the virus will make its way into Central New York. Onondaga County already battles with EEE and West

Nile viruses. Swamp areas in Onondaga County are monitored, and spraying for mosquitos, while protested by some, holds the threat to a minimum. The North Atlantic is no longer a barrier to Old World diseases migrating to the New World. We will see more epidemics. Hopefully, the lessons from the epidemics we have suffered will stand us in good stead when they appear.

17

FALLING FROM THE SKY

Aircraft Crashes

Wilbur and Orville Wright made motorized heavier-than-air flight possible. We had been gliding before then. Men like Otto Lillienthal took to the skies in engineless gliders, and some fell to their doom. Even before Lillienthal rose in motorless splendor into the German skies, the Montgolfier brothers rose in the first hot air balloon. Since man has observed birds in flight, we have longed to share the skies with them. In the 113 years since the brothers Wright lifted into the air at Kitty Hawk, North Carolina, we have seen advances far beyond those Wilbur or Orville could have imagined. In the last days of 1903, cars were still seen as playthings for the rich. In 1908, the military would experiment with aircraft, but many of the army's generals saw the airplane as a toy of no use to modern warfare. That year also saw the first fatality as the result of an aircraft accident. Today, aircraft can circumnavigate the globe in a matter of hours. As we took to the skies with greater frequency and ease, we also learned that gravity can be unforgiving. While air travel remains one of the safest ways to travel, it is aircraft accidents that garner front-page headlines and lead story status in the news. There is a reason they call Captain Sullenberger's landing "the miracle on the Hudson."

In the 113 years since the Wright brothers first lifted off the ground near Kitty Hawk, North Carolina, aviation has cut travel times dramatically. Onondaga County has been lucky. We have not suffered the crash of an airliner as have some areas. We have had our share of military and private airplane crashes.

Baldwinsville, in March 1955, was a sleepy farming community. By the mid-1950s, aircraft were no longer playthings of the wealthy. While still expensive, aircraft could be purchased by the middle class if they were surplus or secondhand. Farmers Allen Barnes and Frederick Woods were killed when their light aircraft plunged to earth. The mishap occurred when the pilot attempted to turn the aircraft to make a tailwind turn and put the airplane into a spin. The aircraft gouged a two-foot hole in a farm pasture on Sorrel Road. Investigators could not determine who was piloting the stricken craft, as it was equipped with dual controls. Barnes was thrown from the plane. The county coroner ruled the deaths accidental.

Military flying differs from civilian flying. All pilots need to train. Private pilots and commercial pilots train to get from point A to point B as safely as possible in the aircraft they fly. While this is part of military flying, there is the added factor that military airmen also train to fight their aircraft. That is, they train to effectively use their aircraft as a weapons platform. On Saturday, July 23, 1955, Air Force lieutenant Everett Taylor took off on a training mission from Hancock Air Force Base in Mattydale. During his flight, the engine of his F-94 Starfire flamed out. Taylor wrestled the jet away from the hamlet of Mattydale and managed to crash the fighter in a Pitcher Hill oat field just west of what is now the Northern Lights traffic circle. Taylor survived his injuries. No one died in this incident thanks to Taylor's skillful stick work. It could have been a devastating tragedy. Had the jet fighter crashed in Mattydale, there would have been a significant body count. Immediately following World War II, Mattydale experienced a housing boom, as returning veterans bought homes under the recently enacted G.I. Bill. Mattydale transformed from a rural farming community to a middle-class suburb. Hundreds of people could have been killed, injured or displaced from their homes. Lieutenant Taylor could have bailed out of the stricken fighter; instead, he chose to ride the plane to the ground and do all he could to prevent innocent casualties. If he wasn't awarded the Distinguished Flying Cross for his actions, he should have been.[58]

Flying can be hazardous in the best of weather. We expect our military pilots to be able to fly in all sorts of weather, and they are trained to do so. Lieutenant John Kessel was participating in an air defense training mission on November 26, 1955. Kessel and his wingman, Captain John Etherington, encountered a snow squall south of Baldwinsville. According to Etherington, Kessel became disoriented, and his jet struck Onondaga Lake and sank in about sixty feet of water. Captain Etherington later described the mission as "another routine mission." Portions of the aircraft seat, life raft and wing tip fuel tank were recovered by navy divers from the naval reserve center in Liverpool. At this writing, I believe the remnants of Lieutenant Kessel's Lockheed Starfire remain at the bottom of Onondaga Lake.[59]

By 1956, the first generation of jet aircraft had been taken out of service or relegated to training roles. The Lockheed F-80 Shooting Star was developed in 1943, and two were seconded to the Eighth Air Force in April 1945 but saw no combat in the waning days of the war. When the Korean War began in 1950, the now venerable F-80 was relegated to the ground attack role, a task it proved well suited to. In spite of its flaws as a fighter, the Shooting Star shot down the first aircraft in the Korean War. It was on these bones that the T-33 trainer was built. The T-33, also known as the Shooting Star, was one of the most widely used jet trainers in the world.[60] Military pilots in the years since Wilbur and Orville first took to the skies learned some hard lessons—notably, that different aircraft respond differently. As aviation technology moved forward and machines became more complex, pilots could no longer teach themselves how to fly. From flight school, military pilots go to training squadrons, where they learn the complexities of modern military aircraft. As it is today, it was the same in 1956. Pilots fly training missions to maintain and increase their proficiency. Geddes native Lieutenant Gordon Simonds was flying just

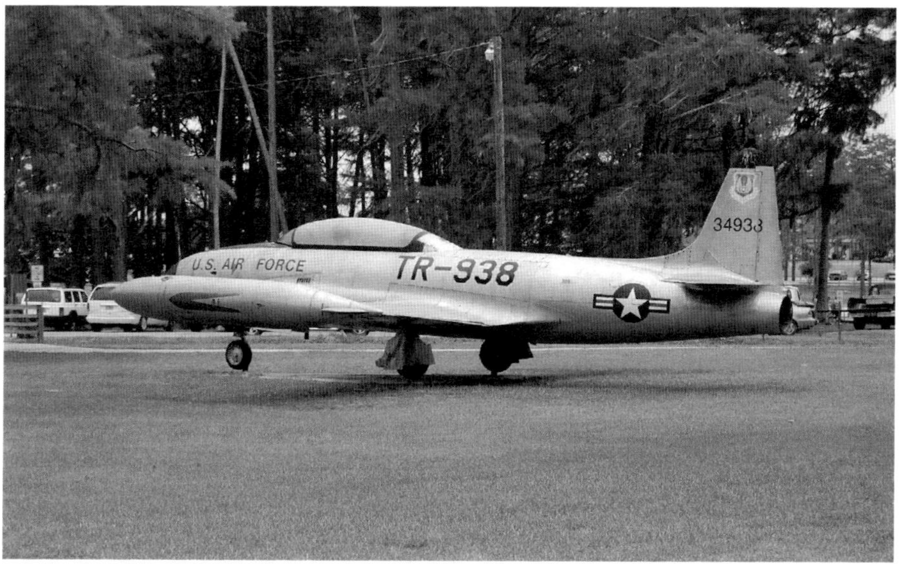

A T-33 Shooting Star similar to the craft that crashed in April 1956 in Skaneateles Lake, killing its pilot. *Courtesy of Peggy Reilly.*

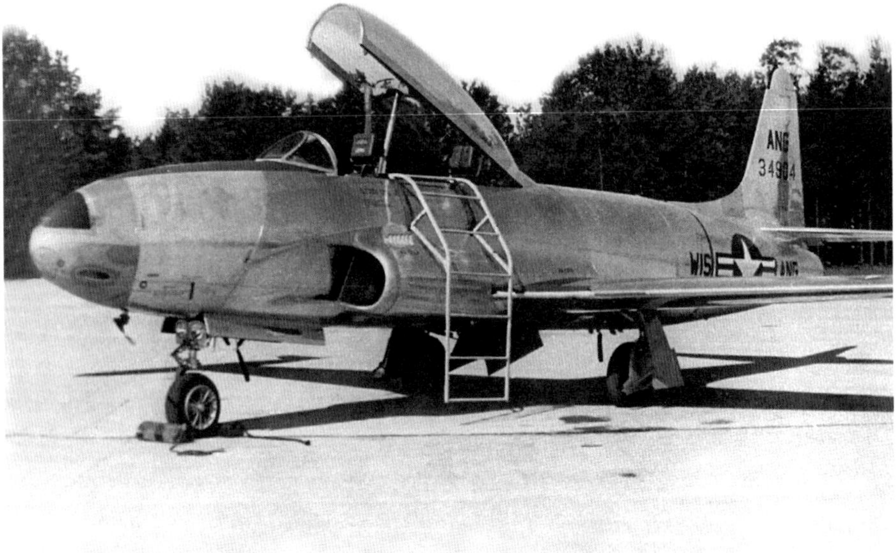

A Lockheed F-94 Starfire. The fighter is similar to the one that crashed into Onondaga Lake in November 1955. *Courtesy of the Library of Congress.*

such a training mission on April 28, 1956. Around 2:30 p.m., his jet went through a series of rolls and barreled into Skaneateles Lake. According to eyewitness accounts, Simonds tried to right the stricken aircraft up to the moment when it plunged into the lake. Some witnesses reported seeing debris fly off the T-33 before it began its fatal last moments. Both mechanical malfunction and pilot error could have been factors in the crash. As in the Pitcher Hill crash, the results could have been so much worse. We will likely never know exactly what Simonds thought in those last moments. He was trying to recover a situation that had gone horribly wrong and perhaps attempting to ensure his stricken craft didn't kill anyone in the town of Skaneateles. Sixty years have passed since the day Lieutenant Simonds died, and we don't know what contributions the young Air Force Reserve pilot might have made had he lived. I prefer to think he died doing his best to avert a major catastrophe while in the service of his community and his nation.[61]

Major Howard Froman had split seconds to make a life or death decision. We like to think we make these decisions, and sometimes we're right. The

Mattydale resident could choose to dump the wing tanks of his jet and eject after the T-33 suffered a power loss, or he could attempt to ride the aircraft in and hope for the best. Releasing the fuel tanks and ejecting could have set houses in Mattydale on fire and killed scores of people. There was never a choice for the veteran U.S. Air Force pilot. He stayed with the damaged airplane and began a long glide to Hancock Field. The Shooting Star was heavily laden with fuel and landed twenty-five feet short of the runway. The jet skidded along the runway, shearing off parts of the landing gear and causing extensive damage to the aircraft, but the wing tanks were intact when the jet came to a stop. Other pilots who observed Froman's landing praised his skill and his courage in staying with the aircraft. Again a major disaster was averted by a pilot's split-second decision.[62]

Military pilots generally spend hundreds more hours in the cockpit than civilian pilots do. Both are liable to suffer mishaps in the air. Four people—Donald Monnier, Bruce Hutchinson, Sandra Lee Story and Donna Hutchinson—died when the Cessna Skyhawk they were flying in struck high tension wires while attempting to land at Hancock Field. The aircraft struck wires near Northern Boulevard in the town of Dewitt. The pilot was attempting night landing when the aircraft struck the power lines. Three of the plane's occupants were pronounced dead at the crash site. The fourth died while in route to the hospital. Approximately two to three thousand homes were without power because of the incident. The pilot had requested clearance to land from the tower, and there were no indications that the Cessna was suffering any sort of mechanical problem. The victims were all in their late teens and early twenties.

Two people died and a third faced charges after their aircraft crashed in Lysander. The lone survivor of the crash and David O'Donnell shared the front seat, and Paul Hyland was in the rear seat when the aircraft landed short of Wikstrom Field. The aircraft propeller dug up earth before flipping, and the aircraft bounced once as it landed. The throttle was wide open, and the crash tore the landing gear from the plane. The engine was found thirty feet from the fuselage. Officials of the Civil Aeronautics Administration, the fore runner of today's FAA, stated that Murphy, who piloted the craft and was the sole survivor, faced charges of operating an unlicensed aircraft and operating an overloaded aircraft. The plane was rated and equipped for only two occupants. The May 23 incident came less than a month after the crash that took Lieutenant Simonds's life.[63]

Thirty-one years later, four people died in a Thanksgiving Day plane crash near Skaneateles Lake. Alan Keith was making his final approach to the

Skaneateles Aerodrome when the Cessna 210 he was flying slammed into the west bank of Skaneateles Lake about two miles from the village. Keith had been flying on instruments and was a veteran pilot with over twenty years of flying experience. Although there was low cloud cover at the time, there was no fog according to other pilots who witnessed the crash. Keith flew out of the Skaneateles Aerodrome frequently and was familiar with the facility. According to an air traffic controller who declined to be named for the newspaper account of the crash, there was no indication of a problem. The Cessna burst into flames on impact. Keith's wife, Virginia; a family friend, Wanda Zobkiw; and Zobkiw's daughter Lisa were passengers on the flight.[64] It is easy to second guess after the fact. Keith was an experienced and capable pilot who got caught up in a catastrophic whirlwind of events at the worst possible time.

A pilot walked away from an aircraft crash in Cicero. The pilot took off from Michael Field on Route 11 in the town of Cicero when his Piper tri-pacer lost power. The pilot landed in a grassy field near the runway. Officials credited the weedy field with cushioning the aircraft.

A Cazenovia man was killed April 18, 2008, when his amphibious plane crashed. His flight instructor was seriously injured. The instructor stated they were simulating an engine failure during landing when the Lake LA-250 lost speed. The plane crashed into a field and skidded into the tree line, where a pine tree fell on the craft. Witnesses reported hearing the engine sputter just before the plane crashed.

As long as we strive to touch the skies, we will live with the risk of aircraft crashes. As I stated earlier, Onondaga County has been lucky in that it has never suffered a major aircraft crash in which an air liner crashed, killing hundreds. Some will question whether the accounts I list here constitute disasters. I would counter the loved ones of the deceased likely saw them as disasters. The other cases I listed are examples of what could have happened but for the bravery and quick thinking of selfless pilots making the ultimate sacrifice or being willing to, in order to avoid a dramatically worse fate for scores of people. Things will fall from the skies over Onondaga County. It is hoped that no more aircraft will.

18
WHAT HAVE WE LEARNED?

We are near the end of our journey, you and I, and I wonder what we have learned? Each catastrophe we endure teaches us lessons, and sometimes we actually heed them. Some of the disasters I mention may not seem like a big deal to some of you, and I get that. I ask that you put yourself in the position of the victims. As the county grew, the potential for truly devastating events increased exponentially. Some of the disasters that scourged the county were compliments of Mother Nature. The others were caused by our fellow human beings. In some cases, our actions were abysmal, but in most, Onondaga County residents rose to their higher natures and pulled together to overcome the adversities thrown at them and their fellow residents.

Here in Onondaga County we're lucky when it comes to natural calamities. We don't often see hurricanes or, for that matter, tornadoes. In all my years living in the Syracuse area, I can remember but one earthquake. It was centered near Albany, and it didn't even move the knickknacks. Hurricane Hazel visited and provided my parents and grandparents with a lot of fodder for the stories they told afterward. The tornado of 1912 cut a swath of destruction through the northeastern section of the county and left a few bodies in its wake. Eighty-six years later, the Labor Day Storm raised havoc with the county.

We get snow here. Syracuse and Buffalo vie with each other for the Golden Snowball award. The award—and to some of us it is a dubious distinction—is given to the city with the greatest snowfall for the season. Snowfall amounts that cripple cities farther south are par for the course here. I spoke several years ago to a gentleman from Maine who was in awe of the snowfall amounts we receive. Our annual snowfall average is approximately ten feet a year. My grandparents spoke of the blizzard of 1925, and I tell my nieces and nephews about the blizzard of 1966. I expect they'll tell their kids about the blizzard of 1993.

Snow melts and rain cause flooding. Normally it isn't too bad. Some years, in parts of the county, the flooding is epic. Onondaga County fought a long, drawn-out battle to control the annual flooding of Onondaga Creek. Some homes on the southern shore of Oneida Lake deal with the same thing. The village of Tully in the southern part of the county had a destructive mudslide. Houses were destroyed, and people were displaced. Fortunately, no one was killed. When the land was cleared and plowed, there was nothing to hold the soil in place when it became saturated.

On the whole, we've been fortunate when it comes to natural disasters. Onondaga County is blessed with a benign climate, and it is not on any major fault lines. I don't know what sort of havoc climate change will play on the county in the years to come. This year was the first I remember in which the county was considered in a drought. Mother Nature isn't the problem usually when it comes to disaster in Onondaga County.

We are our own worst enemies. Accidents happen, but there is an old adage that says there is no such thing as an accident. Most of the man-made calamities Onondaga County has faced are the result of human error. You'll notice I didn't mention epidemics with natural disasters. Humans may not have invented cholera, smallpox, influenza, polio or AIDS, but they helped, even if unwittingly, to spread them. In the days before indoor plumbing, sanitation was sketchy, and the wonder isn't how many epidemics happened in those days, but rather that so many more didn't happen. With the advent of indoor plumbing, diseases like cholera disappeared from Onondaga County. The Spanish flu ravaged those who in other cases would have been the most likely to weather the storm. The disease swept through the young with particular savagery. Polio targeted young children, and my parents' generation rightly viewed Dr. Sabin as a hero. AIDS was a whole other ball of wax. As I stated, no disease is more misunderstood or more prone to feature in conspiracy theories. It would be easy to believe that racial and sexual bigotry have kept funding from the search for a cure. It is one of the few diseases people refuse to talk about. The opioid and heroin epidemic is proof that the war on drugs is by no means won. It is another case of stigmatizing the victim instead of searching for a real solution to the problem. The good news is that better preventative measures, better treatment and willingness to talk about these diseases and how to avoid them is more commonplace these days. Perhaps we will get to the point where we can discuss diseases with no hint of recrimination and hindrance to their cure.

We have seen fires in our history and will continue to do so. The men and women who fight them in most of the county are unpaid volunteers who risk life and limb to save their neighbors' property and lives. Syracuse has a paid professional fire department, but it, too, doesn't do it for the money. Several have died in the call of duty, and they are honored by the firefighter's monument in Fayette Park. Eight of them died in the Collins Block fire of 1939. County landmarks were lost to inferno. We've gotten better at preventing fires in the home and at our places of business. Our fire departments train to continue to improve their skills in combating mankind's oldest scourge. As long as we erect structures, there will be fires and there will be stalwarts who will willingly rush into those infernos to quell them and save those affected by the flames.

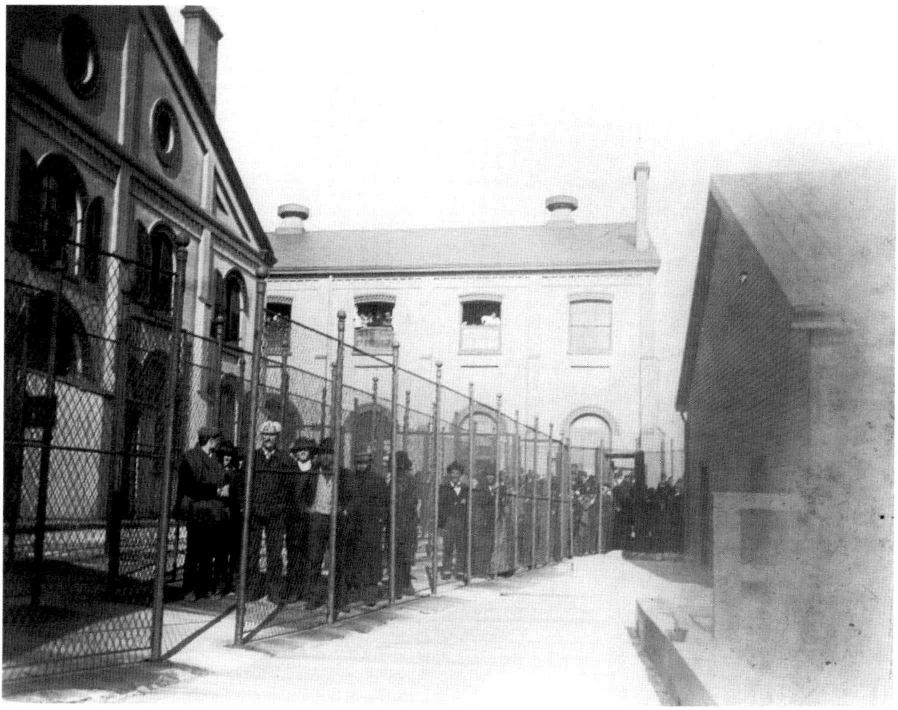

Immigrants in quarantine at the start of the twentieth century. Quarantine was one of the few effective ways to stop the spread of disease, and it was common practice to quarantine immigrants in the waning days of the nineteenth and early days of the twentieth centuries to prevent outbreaks, such as the cholera epidemics of the 1850s. *Courtesy of the Library of Congress.*

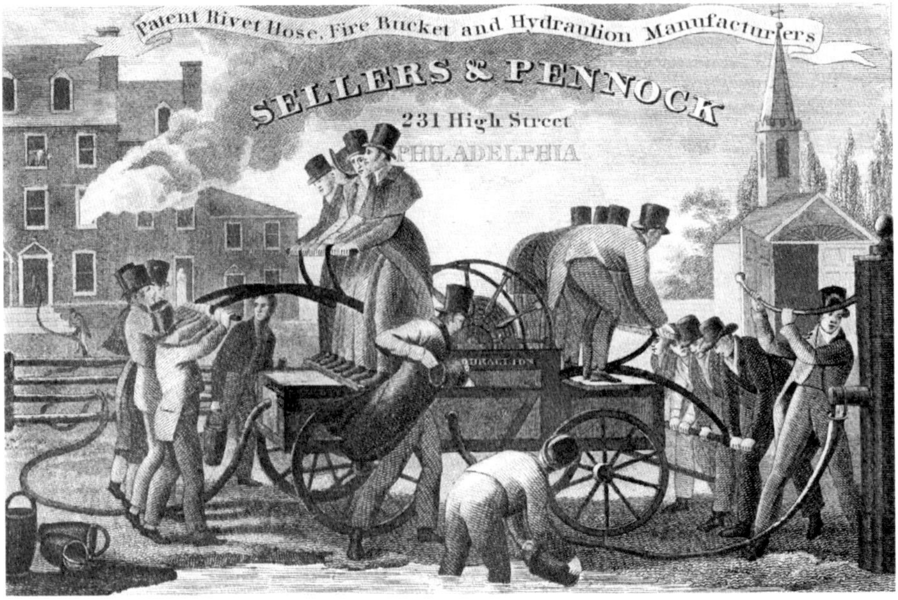

An advertisement for a hand pumper common to the mid-1800s. *Courtesy of the Library of Congress.*

There were steam explosions that killed and maimed people. A family was decimated when the yacht *Trilby* suffered a boiler explosion. It wasn't the first boiler explosion; the steamer *John Greenway* was ruined in 1885 with loss of life. In the case of the *Trilby* a flaw in the boiler at the time of construction caused the blast. The *Greenway*'s demise was brought about by the captain who hired his nephew as engineer in spite of the fact the young man was not qualified, and he ignored warnings that the vessel was not seaworthy. His hubris cost both men their lives. Both the great gunpowder explosion of 1841 and the Split Rock explosion of 1918 still offer up faint whiffs of conspiracy. Why was Charles Goings storing so much gunpowder in his joinery that the blast was felt in Fulton thirty miles away, and what was the enigmatic Mr. Ingalls's role in the affair? We may never know. There was no such thing as forensics in 1841. Seventy-seven years later, the Semet-Solvay munitions plant at Split Rock Quarry would be leveled by an explosion that claimed the lives of up to seventy men. Rumors of sabotage surrounded the investigation. The Onondaga County district attorney ruled the cause was accidental. The explosion was caused by an overheated gearbox in a chemical mixer, but questions remain in both cases. We may never know the answers in either case unless "new" evidence is found in dusty archives or hidden in someone's attic. Neither scenario is likely.

We saw the brutality of war, first in Goose Van Schaik's raid on the Onondaga and centuries later when terrorists bombed Pan-Am Flight 103. We deal with the threat on a daily basis. As I type this, there is a possibility that Al Qaeda will attempt a strike. We live with ISIS and Al Qaeda's threats, and we endure.

We erect buildings for business, domicile and worship in Onondaga County. The vast majority of contractors are reputable and endeavor to construct high-quality edifices that will stand for decades. Sometimes, people cut corners—such was the case with the Central Baptist Church. Fourteen children died and one hundred were injured due to cutting corners to save money. It is a story that bears heeding because it continues to this day. Eighteen years later, the Greenway Brewing Company would lose thousands of dollars in material and two employees would be injured when a portion of the fifth floor of the sprawling brewery collapsed, taking the lower floors with it. In 1907, the building's back wall collapsed when the Erie Canal washout occurred. The brewery moved after the building burned to the ground.

The bottom dropped out of the Erie Canal on July 30, 1907. Vessels were destroyed and could not be lifted from the now-empty canal. They were burned in place. Several buildings were damaged, but astonishingly, no one was killed. The Erie Canal faced steep competition from the railroads almost from its beginning. When automobiles became affordable for businesses and the individual consumer, the canal was no longer viable. Sixteen years after the collapse, the transformed canal debuted as Erie Boulevard. Sections of the canal are now parks with walking trails. If the Erie Canal is gone in Onondaga County, it still casts a huge shadow. The canal is not only integral to New York and Onondaga County history, it is legendary. Stories are still told of canaullers, and ghostly tales surrounded the canal and the communities that populated its banks.

We travel faster than Ephraim Webster and Harvey Baldwin ever dreamed was possible. With speed and population growth came the potential for horrifying and gruesome accidents. Trains and their smaller cousins, the trollies, carried people at what seemed incredible speed from Point A to Point B. Sometimes, they derailed. Other times they hit other trains or people. One man died when he fell from a horse-drawn trolley. He was riding on the roof of the trolley when it collided with another. In the wreck, he was thrown from the roof and died as a result.

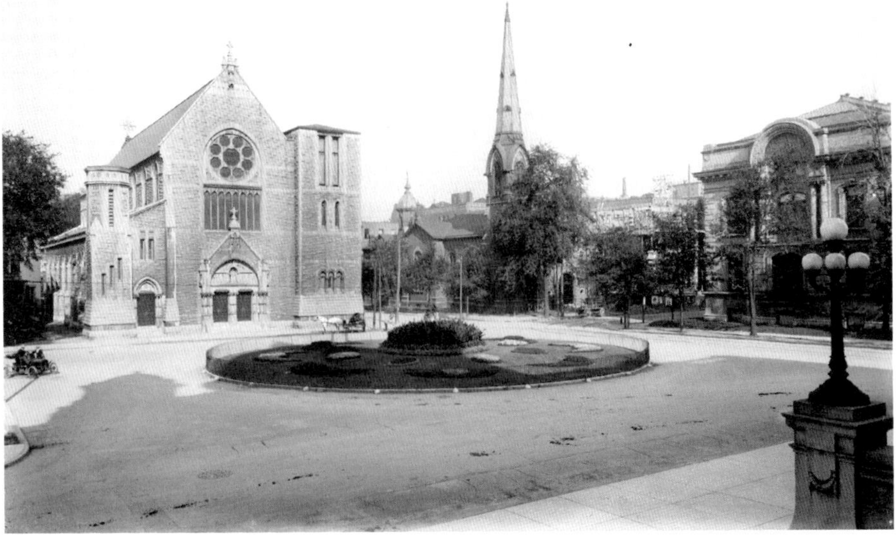

In 1874, Columbus Circle was a center for houses of worship. *Courtesy of the Library of Congress.*

Trains derailed and will continue to do so because of excessive speed. The latest occurrence in Onondaga County happened in Lyncourt. The engineer was going over twice the stated speed for that section when he slammed into another train. People have died over the years in these mishaps. The new positive braking control system will help and has in the areas where it has been installed. We don't suffer incidents with trollies any more. The trolley, like the stagecoach, is a relic of a bygone era. The bus has replaced it, and Onondaga County has seen very few accidents with motor coaches.

The twentieth century saw great improvement in Onondaga County's roads. In the 1930s, the county floated the idea of Onondaga Lake Parkway. The original plan was to route it through Galeville where Route 57, Old Liverpool Road, is, but the state kyboshed the idea because drivers would not be able to see the lake. The parkway runs along a disused portion of the Oswego Canal. It also passes under one of the lowest, if not the lowest, railroad overpasses in the county. Despite numerous signs and traffic devices, several trucks and a bus have run into the overpass. It took almost seventy years before there were fatalities. Unfortunately, all of these accidents and the four fatalities did not have to happen. At the time of the fatal bus crash in 2010, there were over sixteen signs warning of the bridge and stating commercial traffic is banned from Onondaga Lake Parkway. As sure as we will see snow in Onondaga County, we will see someone run into the overpass. Hopefully, there will be no injuries except to the driver's pride and his or her wallet.

As man took to the skies, we found that gravity is unforgiving. We have been fortunate that Onondaga County never saw a commercial liner go down as some cities have. Military and private planes have fallen to earth. To a man, the military pilots strove to ensure their craft would not strike populated areas. We don't know about the civilian pilots. Without knowing, I choose to give them the benefit of the doubt. These days, we see drones in the skies along with manned aircraft. The azure reaches are getting a wee bit crowded.

From the earliest days, we strove to have the progress industry afforded the county. It was a double-edged sword. For years, Onondaga Lake has been noted to be one of the most polluted lakes in the country. The situation is seeing improvement, but decades of sewage and industrial waste have wreaked havoc on the lake's fragile ecosystem. Some of the waste came from the Solvay Process Company and those who took the helm. That the corporation did much good in the county is undeniable. It is equally undeniable that the chemicals that ended up in the lake will take years to clean up. It is possible it may never be totally mitigated.

It is impossible to overstate the great changes wrought by the advent of electricity. It is almost as hard to overstate the effect massive blackouts have on our society. We've seen two major blackouts not caused by storms in Onondaga County that I remember. The first, the great Northeast Blackout of 1965, was caused by a cascading failure. Here in Onondaga County it didn't last terribly long, but it became legendary in myth and movie. The blackout of 2003 was caused by a computer glitch. While our power grid has improved, there are still vulnerabilities. Thirteen years after the 2003 blackout, computer-controlled systems are prone to the efforts of hackers. Some are pranksters. Others are intent on destroying our infrastructure and our way of life. Both have the potential to cause epic calamity. It is a vulnerability we need to address.

With all the uncertainty, there is one constant and that is the people of Onondaga County. We may not always get it right. We will see our share of mishaps and, yes, disasters. Some of us will be the cause of them. We will be at our worst. Yet we will rise to the occasion. There will be firefighters who will enter burning and collapsed structures on the rumor that someone's inside. There will be ordinary people who will perform extraordinary feats to help their stricken neighbors. We will learn from our calamities. We have done so since our earliest days. We will joke about the snow and every other catastrophe that comes our way.

If you take nothing else from this look at our history, know that this is not a glorification of disaster. It is, rather, a tribute to those who overcame adversity. There is little speculation here. These things happened. Our journey has ended. I hope it was an informative and fun one for you. It was for me.

NOTES

Chapter 1

1. Geoff Herbert, "Syracuse Named Safest U.S. City for Refuge from Natural Disasters," *Syracuse*, last updated August 16, 2013, http://www.syracuse.com/news/index.ssf/2013/08/syracuse_safe_cities_natural_disasters_refuge.html; John Metcalfe, "These Cities Are the Safest Refuges from Natural Disasters," *Atlantic Monthly* (August 2013).
2. The average snowfall for Syracuse is approximately 120 inches per year, or 10 feet.
3. *Webster's Ninth New Collegiate Dictionary* (Springfield, MA: Merriam-Webster, 1984).

Chapter 2

4. Readers will glean where the chapter title comes from.
5. Baum was from the town of Chittenango in neighboring Madison County. His wife was the daughter of prominent suffragist Matilda Joslin Gage, who was from Fayetteville in Onondaga County.
6. The hamlet of Longbranch was absorbed into the town of Salina. There are several hamlets and villages that are parts of larger towns in Onondaga County. Longbranch is still a park in the county park system, but it has disappeared as a separate governmental and town entity.
7. Paul J. Hebert and Glenn Taylor, "The Deadliest, Costliest and Most

Intense Hurricanes in the United States of This Century (and Other Frequently Requested Hurricane Facts," National Weather Service, January 1983, http://www.nhc.noaa.gov/pdf/NWS-NHC-1983-18.pdf.

8. Ibid.
9. Other names have been retired since, including Katrina, Sandy and Andrew. The names are associated with massive destruction and loss of life, and their use could seem insensitive.
10. Syracuse was a major center for the manufacture of religious candles in the twentieth century. Today, that industry has moved elsewhere.
11. "The New York State Labor Day Derechos," *Wikipedia*, last modified February 17, 2017, https://en.wikipedia.org/wiki/New_York_State_Labor_Day_derechos.
12. My wife and I were without power for three days. My mother-in-law, who lived a scant three miles from us, never lost power.

Chapter 3

13. Henry W. Schramm and William F. Roseboom, *Syracuse from Salt to Satellite: A Pictorial History* (Woodland Hills, CA: Windsor Publications, 1979), 25–27.
14. Initial accounts stated fifteen dead, but as many as twenty-five may have died in the blaze.
15. Though as the name implies, operas were staged in these venues, the term *opera house* was adopted for these multiuse venues to give an air of culture and sophistication. The Earlville Opera House in neighboring Madison County still operates.
16. *New York Times*, October 6, 1862.
17. Ironically, Keep's husband was the president of the Typographers Local.

Chapter 4

18. Although the exact location of the Onondaga village is unknown, the approximate location is near the corner of West Seneca Turnpike and Midland Avenue. The exact location is hard to pinpoint because the path of Onondaga Creek was altered to alleviate flooding in the twentieth century.
19. Webster's Square Mile is in the Valley and encompasses Webster's Pond on Valley Drive, a popular attraction for local bird lovers and anglers.

20. America had lived through hijackings and other terrorist acts before Pan-Am 103 and, of course, 9/11.
21. Sandy Grady, "U.S. Sounded Weak Warning," *Philadelphia Daily News*, December 25, 1988.
22. The flight originated in Germany, and passengers transferred from a Boeing 727 to the larger 747 in London.
23. Unlike American or British jurisprudence, which only has two verdicts, guilty or not guilty, Scottish law has three: guilty, not guilty and not proven.

Chapter 5

24. The dead were John Austin, Dr. O.E. Wainwright, Charles D. Collins (aged fourteen), Mrs. E.A. Barnes, Augusta W. Carpenter, Minnie Collins, Mrs. James Crow, Grace Holmes, Mrs. J.E. Karr, Mrs. E.A. Vedder, Carrie Ostrander, Harriet Leonard, Lulu Horton and Minnie Thomas.
25. There is dispute as to what the band played. Most sources state it was the hymn, but others equally credible state that the band was playing the popular song "Autumn." Both story versions come from survivors.
26. *New Haven Evening Register*, January 6, 1880.
27. I will cover the collapse in fuller detail in a subsequent chapter.

Chapter 6

28. The name is as spelled in contemporary accounts, though his name may have been Eugene.
29. A rod is sixteen and a half feet and is no longer used as a measurement.

Chapter 8

30. It has been said of Chicago that a politician can do anything they want as long as the streets are kept clear of snow.
31. The city of Oswego on Lake Ontario, approximately forty-five miles north of Syracuse, garnered 103 inches of snow during the blizzard.

Chapter 9

32. While not noted for earthquakes, Upstate New York has suffered them on occasion, most recently a couple of years ago, resulting in no damage.
33. Church Street is now called East Willow Street.
34. Neil K. MacMillan, *Haunted Onondaga County* (Charleston, SC: The History Press, 2015), 32–35.
35. The opposing sides were the Triple Entente: France, Britain and Russia—later called the Allies—and the Central Powers: Germany, Austria-Hungary and the Ottoman Empire.
36. Many accounts put the death toll at fifty; however, the *Syracuse Herald* for July 5, 1918, stated seventy died in the blast. The *Herald* of July 9, 1918, claimed sixty-one dead but added that several of the injured were not expected to survive their injuries.
37. Iva Spink's name is also spelled Ivas in an article by the *New York Times* about the event.

Chapter 11

38. I assume the 1838 culvert replaced another structure for the same purpose, though I have not found documentation to support that assertion.
39. Sean Kirst, "Hidden Treasure Restored in Downtown Syracuse: The 19th Century Erie Canal Bridge Few Ever See," *Syracuse*, last modified October 30, 2014, http://www.syracuse.com/kirst/index.ssf/2014/10/in_syracuse_the_most_beautiful_erie_canal_bridge_no_one_ever_sees.html.
40. Ibid.
41. The year 1907 was an active one for disaster in Onondaga County. In addition to the canal collapse, there were two fires that rocked the county (see chapter 3).

Chapter 12

42. *Syracuse Daily Journal*, July 25, 1860.

Chapter 13

43. The Labor Day storm, which I cover in another chapter, is a notable exception.
44. The eight states were Connecticut, Massachusetts, New Hampshire, New Jersey, New York, Pennsylvania, Rhode Island and Vermont.
45. "City, County Take Blackout in Stride," *Syracuse Herald-Journal*, November 10, 1965.
46. Ibid.
47. Dennis William Hauck, *Haunted Places: The National Directory* (New York: Penguin Books, 2002), 305.
48. "City, County Take Blackout in Stride."

Chapter 14

49. Mark Weiner, "Mudslide Puts on a Rare Show," *Syracuse Herald-Journal*, April 28, 1993.

Chapter 15

50. Delen Goldberg, "Without a Paddle," *Syracuse Sunday Post-Standard*, July 15, 2007.

Chapter 16

51. G. William Beardslee, "The 1832 Cholera Epidemic in New York State," Varsitytutors.com, 2016, https:// www.varsitytutors.com/early-america/early-america-review,volume-4/the-1832-cholera-epidemic.
52. The primary difference between a pandemic and an epidemic is the area affected. A pandemic usually has continental or global repercussions, while an epidemic is more localized.
53. "Smallpox," *Wikipedia*, last modified February 12, 2017, https://en.wikipedia.org/wiki/Smallpox.
54. Four thousand were soldiers training at Camp Syracuse, located on the state fairgrounds. The remainder were civilians infected as the disease spread.

55. *Syracuse Herald*, August 3, 1916.
56. Bush meat is meat processed from wild animals in areas of Africa where it has become harder to farm or ranch traditional livestock. The theory says that bush meat processors picked up the virus (HIV-1 and HIV-2) through open cuts.
57. Alan R. Cantwell Jr., MD, "AIDS: A Doctor's Note on the Man-Made Theory," *Paranoia Magazine* (Fall 1997); Al Hiddell and Joan D'arc, *The Conspiracy Reader* (New York: Citadel Press, 1998).

Chapter 17

58. Ibid., November 26, 1955.
59. Robert Jackson, *The Encyclopedia of Aircraft* (New York: Amber Books, 2009), 308–9.
60. MacMillan, *Haunted Onondaga County*, 88–91; *Syracuse Herald-American*, April 29, 1956.
61. Ollie Sheremeta, "Jet Crash Lands at Hancock Field," *Syracuse Post-Standard*, January 12, 1957.
62. *Syracuse Herald-Journal*, May 23, 1956.
63. Ibid., November 27, 1987.
64. *Syracuse Post-Standard*, November 27, 1987.

ABOUT THE AUTHOR

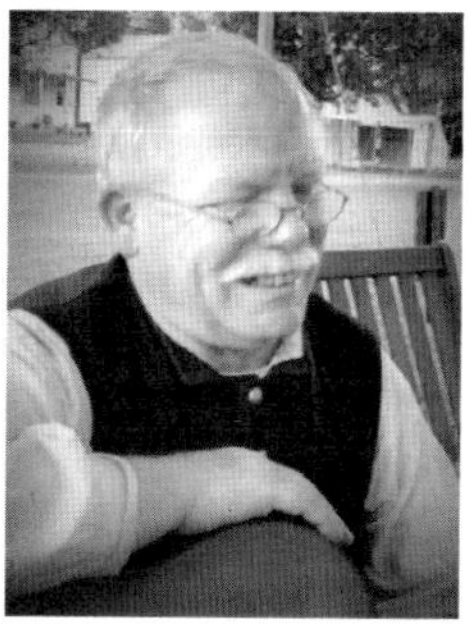

Courtesy of Peggy Reilly.

Neil MacMillan was born and lives in the Syracuse area. He is a U.S. Navy veteran of Operation Desert Shield and Desert Storm and a Vietnam-era veteran. Neil holds a BA in interdisciplinary studies with concentrations in history and creative writing. He has worked in a variety of fields. He is the author of *Wicked Syracuse: A History of Sin in Salt City* and *Haunted Onondaga County* and can be reached a nm@twcny.rr.com.